Veronika R. Meyer

Pitfalls and Errors of HPLC in Pictures

Further Titles for Chromatographers

S. Kromidas (ed.)

HPLC Made to Measure

A Practical Handbook for Optimization

2006. ISBN 3-527-31377-X

S. Kromidas

More Practical Problem Solving in HPLC

2005. ISBN 3-527-31113-0

V. R. Meyer

Practical High-Performance Liquid Chromatography

2004. ISBN 0-470-09378-1

S. Kromidas

Practical Problem Solving in HPLC

2000. ISBN 3-527-29842-8

P. C. Sadek

Troubleshooting HPLC Systems

A Bench Manual

2000. ISBN 0-471-17834-9

U. D. Neue

HPLC Columns

Theory, Technology, and Practice

1997. ISBN 0-471-19037-3

L. R. Snyder, J. J. Kirkland, J. L. Glajch

Practical HPLC Method Development

1997. ISBN 0-471-00703-X

Veronika R. Meyer

Pitfalls and Errors
of HPLC in Pictures

2., revised and enlarged edition

WILEY-VCH Verlag GmbH & Co. KGaA

Dr. Veronika R. Meyer
EMPA St. Gallen
Lerchenfeldstrasse 5
9014 St. Gallen
Switzerland
e-mail: veronika.meyer@empa.ch

Library of Congress Card No.: applied for
A catalogue record for this book is available from the British Library.

Bibliographic information published by Die Deutsche Bibliothek
Die Deutsche Bibliothek lists this publication in the Deutsche Nationalbibliografie; detailed bibliographic data is available in the internet at <http://dnb.ddb.de>.

© 2006 Wiley-VCH Verlag GmbH & Co. KGaA, Weinheim

Printed in the Federal Republic of Germany
Printed on acid-free paper

Cover Design: SCHULZ Grafik-Design, Fußgönheim
Typesetting: Mitterweger & Partner GmbH, Plankstadt
Printing: betz-druck GmbH, Darmstadt
Binding: J. Schäffer GmbH, Grünstadt

ISBN-13: 978-3-527-31372-3
ISBN-10: 3-527-31372-9

1 wort = 1 millibild

(A picture says more than a thousand words)

Graffiti on the Baltzerstrasse in Bern

Preface

Errors are a common companion of all human activity, including work in the laboratory. Yet it is a great pity if erroneous results are produced with great effort and by using expensive instruments and demanding procedures. Therefore a book about sources of errors in high performance liquid chromatography, today's most widely used analytical method, is not superfluous. Maybe the topic is not welcomed enthusiastically but I hope I have found a design which encourages reading and thinking.

In conception, at least, possible problems can be divided into two categories. 'Errors' are troublesome opponents of accurate and precise analytical results which, however, can be understood; we need to remember them frequently. In contrast, 'pitfalls' are totally unexpected intruders and the secret behind them is difficult to discover. The worst are those which are not detected but which affect the result anyway. The book does not, nevertheless, distinguish between the two types. The readers decide how they classify them. With increasing experience in HPLC it should become easier to avoid the pitfalls.

The second edition has been enlarged by the addition of new examples and proposals. Many people have helped me with examples, hints, or ideas on how to improve the text and figures. I want to thank all of them. Special thanks are due to the publisher, who supported the idea of a picture book, not for children but for both novices and experts in the analytical laboratory. I hope that the book will be a useful aid in daily laboratory work thanks to intelligible explanations and lucid illustrations.

St. Gallen, January 2006   Veronika R. Meyer

Table of Contents

Introduction

This book is not an introductory text to HPLC and also not a trouble-shooting guide of the kind "what shall I do if my instrument does not work?". It does not replace such books but is intended to complement them. Some texts which, according to my personal opinion, are very useful and should therefore be present in the HPLC laboratory are listed on the next page.

Now this book on your desk is a picture book. The figures are at least as important as the texts; sometimes more information can be found in them than could be given in the short descriptions. It is possible, and in principle recommended, to study all the pages in sequence from beginning to end. This method guarantees that one learns about errors which are uncommon and unexpected. On the other hand each pair of pages is limited to one topic, linked to other pages by arrows only, and can therefore be studied in isolation. The index at the end of the book can help you to find the right pages when a problem occurs, although it must be stated once again that quick troubleshooting advice is not usually provided.

The book is divided into three parts:

Part I briefly presents some basic facts about HPLC. Many topics may be absent because this is not a textbook, but the matter presented is of utmost relevance in HPLC. Thus the topics discussed should act as reminders and be used for revision. Whoever understands Part I knows a lot about HPLC – more than it seems at first glance.

Part II lists the pitfalls and sources of error. They are in a logical sequence, as far as this is possible, following the flow path in an HPLC instrument, from the preparation of the mobile phase to data evaluation. The list is somewhat arbitrary, and not all errors are of equal importance with regard to their possible consequences. It would, however, be dangerous to distinguish between grave and harmless errors. A minute error can cause much damage under special circumstances.

Part III gives some hints on what can be done to avoid errors. Again this synopsis is very heterogeneous in character. This does not diminish its value, of course.

Incompleteness is an inevitable feature of this book. I am grateful for all hints on other pitfalls and sources of error or on how to avoid them.

Recommended Texts

Veronika R. Meyer
Practical High Performance Liquid Chromatography
Wiley, Chichester
4th edition 2004

John W. Dolan and Lloyd R. Snyder
Troubleshooting LC Systems
Aster, Chester
1989

Paul C. Sadek
Troubleshooting HPLC Systems: A Bench Manual
Wiley-Interscience, New York
2000

Stavros Kromidas
Practical Problem Solving in HPLC
Wiley-VCH, Weinheim
2000

Stavros Kromidas
More Practical Problem Solving in HPLC
Wiley-VCH, Weinheim
2004

Lloyd R. Snyder, Joseph J. Kirkland and Joseph L. Glajch
Practical HPLC Method Development
Wiley-Interscience, New York
2nd edition 1997

Norman Dyson
Chromatographic Integration Methods
Royal Society of Chemistry, London
2nd edition 1998

Part I

Fundamentals

1.1 Chromatography

In chromatography, a physical separation method, the components of a mixture are partitioned between two phases. One of the phases stays in its place and is called the stationary phase, whereas the other moves in a definite direction and is called the mobile phase.

According to the type of mobile phase we distinguish between gas chromatography, supercritical fluid chromatography, and liquid chromatography.

The separation is based upon the different partition coefficients of the sample components between the two phases. It is helpful to divide the chromatographic column into small hypothetical units, the so-called theoretical plates. Within each plate a new partition equilibrium is established. The narrower a theoretical plate, the more equilibrium processes can take place within a column of given length and the more demanding the separation problems which can be solved.

The figure shows the separation of two compounds. One of these prefers the mobile phase but also enters the stationary phase. For the other compound the preference is the other way round. Thanks to this large difference in their properties the two types of molecule can easily be separated. They are transported through the column by the flow of the mobile phase and thereby reach zones where new equilibria are formed again and again.

In the drawing, such a theoretical plate has a height of approximately 3 1/2 stationary phase particle diameters. This height depends on the packing quality of the column, on the mass transfer properties of the phases, and on the sample compounds involved. Plate height is a function of the particle diameter of the stationary phase. For good columns, plate heights are equal to ca. 3 particle diameters irrespective of the particle size. A fine packing, e.g. with a 5-µm phase, gives four times as many theoretical plates as does a 20-µm packing if identical column lengths are compared. The column with the fine packing can therefore be used for more difficult separation problems.

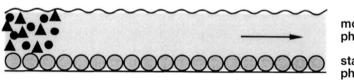

mobile
phase

stationary
phase

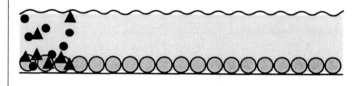

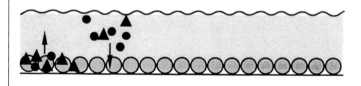

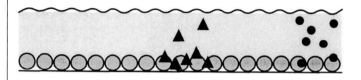

1.2 Chromatographic Figures of Merit

To judge a chromatogram it is necessary to calculate some data which can be easily obtained. The integrator or data system yields the retention times, t_R, and peak widths, w; perhaps it is advisable to determine the peak width at half height, $w_{1/2}$. In addition the breakthrough time or 'dead time', t_0, must be known although it can be a problem to measure it unambiguously. In principle, the first baseline deviation after injection marks t_0. Then the following data can be calculated:

1. **Retention factor, k** (formerly capacity factor, k'):

$$k = \frac{t_R - t_0}{t_0}$$

The retention factor is a measure of the retention of a peak. It depends only on the phase system (the types of mobile and stationary phase) and on the temperature.

2. **Separation factor, α:**

$$\alpha = \frac{k_2}{k_1}$$

Two compounds can be separated only if α is higher than 1.0 in the selected phase system. For HPLC separations α should be 1.05 or higher ($\rightarrow$ 1.3).

3. **Theroetical plate number, N:**

$$N = 16\left(\frac{t_R}{w}\right)^2 = 5.54\left(\frac{t_R}{w_{1/2}}\right)^2 = 2\pi\left(\frac{h_P t_R}{A_P}\right)^2$$

where h_P = peak height and A_P = peak area. The plate number is a measure of the separation performance of a column. (The equations given here are in principle only valid for symmetrical peaks.)

From the plate number it is possible to calculate the height, H, of a theoretical plate (e.g., in μm):

$$H = \frac{L_c}{N}$$

where L_c = column length.

4. **Tailing T** (for asymmetric peaks):

$$T = \frac{b}{a}$$

where a and b are determined at 10% of peak height.

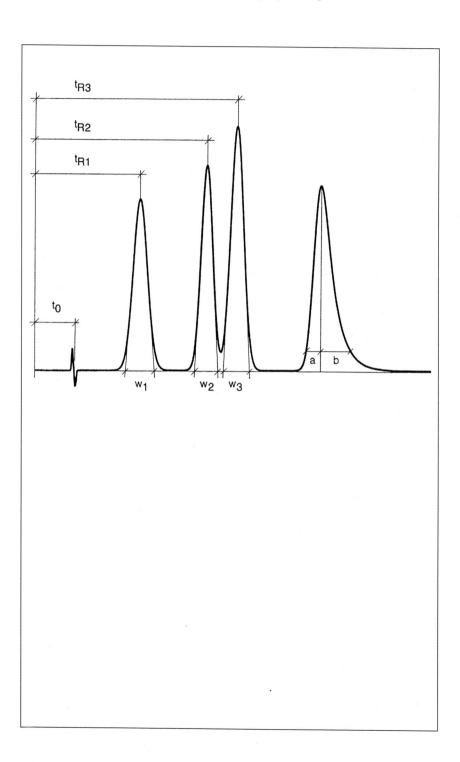

1.3 The Resolution of Two Peaks

The resolution of two adjacent peaks is defined as

$$R = 2\frac{t_{R2} - t_{R1}}{w_1 + w_2} = 1.18\frac{t_{R2} - t_{R1}}{w_{1/2_1} + w_{1/2_2}}$$

At a resolution of 1.0 the baseline between the peaks is not reached! Complete resolution is only obtained at $R = 1.5$ or higher, depending on the height ratio of the peaks. The smaller a peak compared with its large neighbor the greater is the resolution necessary to separate them.

The resolution depends on the separation factor α, the theoretical plate number N, and the retention factor k:

$$R = \frac{1}{4}(\alpha - 1)\sqrt{N}\frac{k}{1 + k}$$

This equation can be expressed in different forms, which are not of interest here. It is important to realize that the resolution is influenced by the three parameters. The separation factor has the largest effect. If a separation needs to be improved it is well worth the effort of increasing α, although it is impossible to give a general proposal concerning how to do this. If the plate number is increased, the effect is only by the factor $\sqrt{N}$; if the column length is, e.g., doubled, and by this also the plate number (at least in principle), the resolution will improve only by $\sqrt{2} = 1.4$. Increasing the retention factor only has a notable influence on resolution if k was small to start with.

The upper figure presents several pairs of peaks separated with varying resolution. The graph below demonstrates how the resolution increases with increasing plate number for three different separation factors.

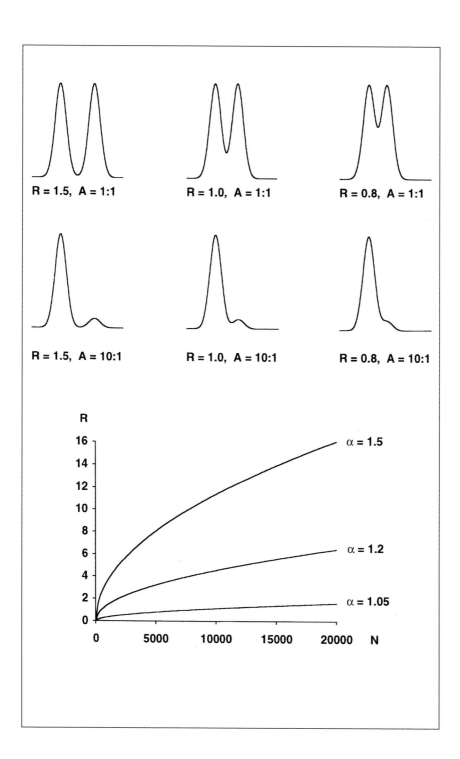

1.4 Reduced Parameters

The judgement and comparison of HPLC columns is best done with reduced, dimensionless parameters. A test chromatogram is acquired which enables the theoretical plate number, N, of the column to be determined from a suitable peak with low tailing. It is also necessary to measure the breakthrough time, t_0, with a refraction index peak or with an otherwise suitable compound (for reversed-phase separations, e.g., with thiourea). The pressure drop, Δp, under the given conditions is noted.

Then the following reduced parameters can be calculated:

1. Reduced plate height, h:

$$h = \frac{H}{d_p} = \frac{L_c}{N \cdot d_p}$$

h is a measure of the height of a theoretical plate as a multiple of the particle diameter, d_p. L_c is column length.

2. Reduced flow velocity, v:

$$v = \frac{u \cdot d_p}{D_m} = \frac{L_c \cdot d_p}{t_0 \cdot D_m}$$

v is a measure of the flow velocity in relation to the particle diameter, d_p, and the diffusion coefficient, D_m. In most cases D_m is not really known but it can be assumed to be $1 \cdot 10^{-9}$ m^2 s^{-1} for small molecules in water/acetonitrile and to $4 \cdot 10^{-9}$ m^2 s^{-1} for small molecules in hexane.

3. Reduced flow resistance, ϕ:

$$\phi = \frac{\Delta p \cdot d_p^2}{L_c \cdot \eta \cdot u} = \frac{\Delta p \cdot d_p^2 \cdot t_0}{L_c^2 \cdot \eta}$$

With ϕ the pressure drop can be described simply and clearly. It is, however, necessary to know the viscosity, η, of the mobile phase. Mixtures of water and organic solvents pass through a maximum of viscosity!

Favorable numbers: $h = 3$
$v = 3$
$\phi = 500$ up to a maximum of 1000

$$h = \frac{H}{d_p} = \frac{L_c}{N \cdot d_p}$$

$$v = \frac{u \cdot d_p}{D_m} = \frac{L_c \cdot d_p}{t_0 \cdot D_m}$$

$$\phi = \frac{\Delta p \cdot d_p^2}{L_c \cdot \eta \cdot u} = \frac{\Delta p \cdot d_p^2 \cdot t_0}{L_c^2 \cdot \eta}$$

$$h = 3$$
$$v = 3$$
$$\phi = 500$$

D_m = diffusion coefficient [m^2 s^{-1}]
d_p = particle diameter [m]
H = height of a theoretical plate [m]
L_c = column lenght [m]
N = theoretical plate number
Δp = pressure drop [N m^{-2}] (1 bar = 10^5 N m^{-2})
t_0 = breakthrough time [s]
u = linear flow velocity [m s^{-1}]
η = viscosity [Ns m^{-2}] (1 mPas = 10^{-3} Ns m^{-2})

1.5 The Van Deemter Curve

The separation performance of a column is not independent of the mobile phase flow rate. An optimum velocity, u_{opt}, is observed where the performance is highest. This relationship is described by the van Deemter curve which describes the height of a theoretical plate, H, as a function of the linear flow velocity, u. At u_{opt} the plate height, H_{min}, is smallest, which means that the number of theoretical plates, $N = L/H$, is largest. The peaks are narrowest and thus eluted with the largest possible height; the resolution reaches a maximum. Any deviation from the van Deemter optimum yields smaller peak heights and resolutions; yet the optimum velocity is not identical for all compounds of a sample mixture.

It would be best to work at u_{opt}. Practical separations are often performed at higher speed, which gives shorter analysis times and usually only a moderate loss of separation performance. This is, however, only true if mass transfer is fast, which is often not the case with special stationary phases and ion exchangers. Of course it is never advisable to work under conditions left of the van Deemter optimum. In this region the separation performance is very poor and the analysis time is long.

If the van Deemter curve is plotted with reduced parameters v and h the optimum is often at $v = 3$.

Chromatographic conditions:

Sample:	5 µL of a solution of thiourea, veratrole, acetophenone and nitrobenzene
Column:	4.6 mm × 7.5 cm
Stationary phase:	Zorbax SB C-18, 3.5 µm (reversed phase C_{18})
Mobile phase:	water/ethanol 3 : 7
Volume flow rate:	0.1 to 1.9 mL min^{-1} corresponding to u from 0.16 to 3.2 mm s^{-1}
Detector:	UV 268 nm
Optimum:	for nitrobenzene (curve) $u \approx 1.3$ mm s^{-1} (0.8 mL min^{-1}), $H \approx 8.5$ µm; for veratrole and acetophenone $u \approx 0.65$ mm s^{-1} (0.4 mL min^{-1})

Reference: V. R. Meyer

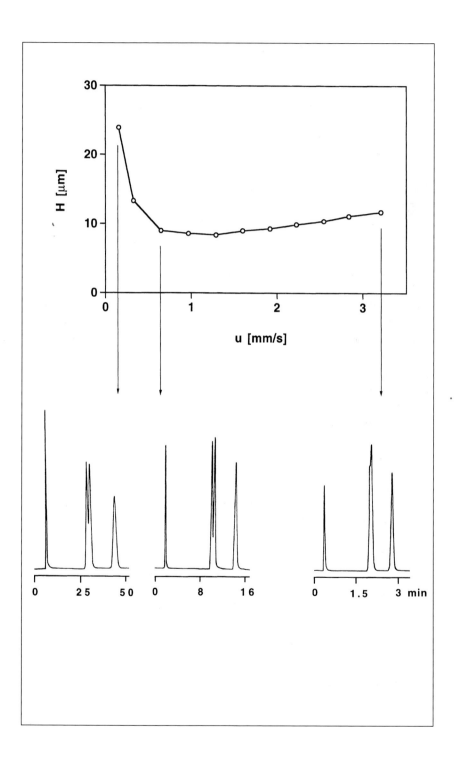

1.6 Peak Capacity and Number of Possible Peaks

For the separation of complex mixtures it is necessary to have space for as many peaks as possible with adequate resolution throughout the chromatogram. This number is the larger the higher the number of theoretical plates of the column and the longer one is willing to wait for the last peak. In other words, N and k determine the so-called peak capacity, n, of the column in use. Usually $R = 1.0$ is taken as the necessary resolution between two adjacent peaks ($\rightarrow 1.3$).

If the plate number were constant over the whole range of k values the peak capacity would be defined as:

$$n = 1 + \frac{\sqrt{N}}{4} \ln(1 + k_{max})$$

At constant peak width, w, as could be the case with a very steep solvent gradient, the peak capacity would be much higher:

$$n = \frac{t_{R\,max} - t_0}{w}$$

In practice neither plate numbers nor peak widths are constant. N increases with time because no chromatographic system has an ideal behavior; and even with gradient elution the peak widths increase gradually. Therefore the Number of Possible Peaks, NPP, is a more realistic value than the peak capacity, n:

$$NPP = \frac{\Delta t_R}{w_n - w_1} \ln \frac{w_n}{w_1}$$

with the retention time interval Δt_R covering peaks 1 to n (including their widths) and peak widths w.

The figure presents a first peak at 2 min, or 120 s, of width 6.2 s ($N = 6000$) and a last at 3.33 min, or 200 s, of width 8.4 s ($N = 9000$). Thus the retention time interval runs from (120-3.1) s to (200+4.2) s or a span of 87.3 s. With these data NPP = 12 as shown in the computer simulation with peaks of resolution 1.0.

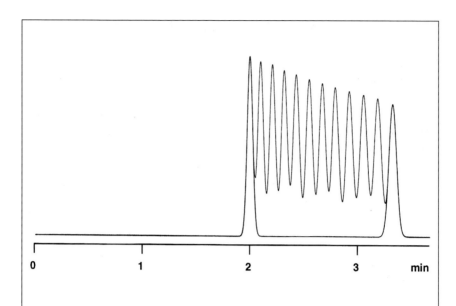

$$NPP = \frac{\Delta t_R}{w_n - w_1} \ln \frac{w_n}{w_1}$$

1.7 Statistical Resolution Probability: Simulation

Even with mixtures of moderate complexity it is not realistic to expect that all compounds can be separated without an extra effort at optimization. The opposite is the general rule – we are forced to assume that there are more compounds than visible peaks. If the separation problem does not deal with homologs or with molecules of another type of regularity, the retention times (or k values) show a random distribution; therefore the chance of peak overlap is high. This statistical resolution probability, P, for a single compound of the mixture is defined by:

$$P \approx e^{-2m/n}$$

where m = true number of compounds in the mixture and n = peak capacity or, for real chromatograms, NPP = Number of Possible Peaks ($\rightarrow$ 1.6).

Example: breakthrough time t_0 = 40 s, maximum retention time = 363 s, thus k_{max} = 8.08 and theoretical plate number N = 5000, giving a peak capacity of n = 40. The sample mixture consists of 10 compounds, i.e. m = 10. The probability that a certain compound will be separated from its nearest neighbor with a resolution R $\geq$ 1 is:

$$P \approx e^{-2 \cdot 10/40} = 0.61 = 61\%$$

This means that we expect that only 6 out of the 10 compounds will be totally resolved! Because this prognosis is based on statistics, it is quite possible that in a given case more or fewer than 6 compounds will have a resolution greater than 1 on both sides.

Computer simulation with random numbers:

Retention time window between 40 and 363 s
Theoretical plate number N = 5000 (constant, less realistic)
Peak capacity n = 40
Number of compounds m = 10
Number of visible peaks = 8, including, however, 1 doublet and 1 triplet
Number of peaks with resolution R < 1 = 5

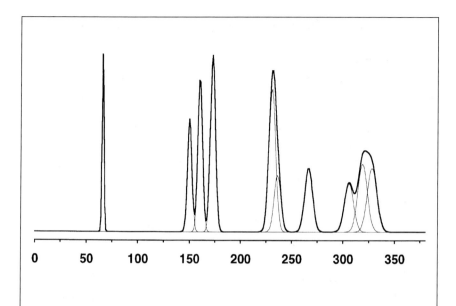

$$P \approx e^{-2\mathrm{m/NPP}}$$

1.8 Statistical Resolution Probability: Example

A mixture of six compounds without chemical similarity needs to be separated. With hexane/tetrahydrofuran (top) and with hexane/dichloromethane (bottom) only five peaks are visible; only hexane/*tert*-butyl methyl ether separates them all (middle). If no extra optimization is tried, in two of the three cases fewer peaks are found than are really present.

   If the chromatogram is recorded by a computer, the data from all the peaks are easily obtained. In the examples shown the theoretical plate number increases from ca. 8000 to ca. 13000 between the first and last peak. The Number of Possible Peaks NPP ($\rightarrow$ 1.6) is, from top to bottom, 19, 27, and 47 (NPP is larger the longer the maximum retention time); the calculated statistical resolution probabilities ($\rightarrow$ 1.7) are 0.54, 0.64, and 0.78. The real resolution probabilities are, from top to bottom, 0.66 (4 of 6 peaks are totally resolved), 1.0 (all peaks are resolved), and 0.33 (only 2 of 6 peaks are resolved; the resolution is 0.9 for peaks 5 and 4).

   The example shows that one possible means of tracking down hidden peaks is to run the sample with several different phase systems. Another method is to use the peak purity function of the diode array detector; this only fails if the unresolved compounds have identical UV spectra or exactly identical retention times.

Chromatographic conditions:

Sample:	1,4-diphenylbutane, 2-phenylethylbromide, phenetole (phenyl ethyl ether), *trans*-chlorostilbene oxide, nitrobenzene and 4-chlorobenzophenone dissolved in hexane and a little THF
Column:	3.2 mm × 25 cm
Stationary phase:	LiChrosorb SI 60, 5 μm (silica)
Mobile phase:	as indicated, 1 mL min^{-1}
Detector:	UV 254 nm

Reference: V. R. Meyer

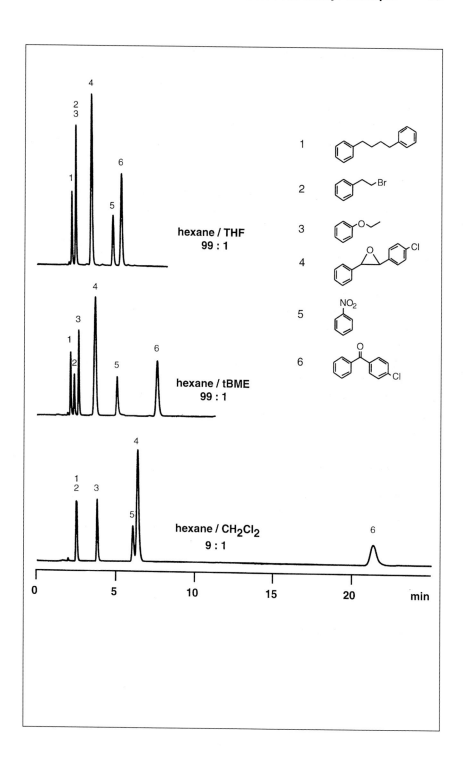

1.9 Precision and Accuracy of an Analytical Result

It is important to distinguish between precision and accuracy whenever an analytical result is evaluated. During method development both aspects must be optimized and it is not acceptable to neglect one of them.

High precision means that the scatter of data is low if the analysis is repeated several times. It is easily calculated as the standard deviation or coefficient of variance of the obtained data (→ 1.10). It describes the random deviations of the method.

High accuracy means that the analytical result deviates only slightly from the true value. It is a measure of the systematic deviations of the method. If the true value is not known, which is usually the case, no statement about accuracy can be made. Knowledge about the accuracy of one's work can be obtained by analysis of reference materials, inter-laboratory tests, and analysis of a particular sample by another method (→ 3.9). Unfortunately, accuracy tends to decrease with decreasing concentration of the analyte in the sample (→ 1.12).

Accuracy is more important than precision. A method with poor precision but good accuracy, as symbolized by the bottom left diagram, needs to be repeated many times but the mean will be accurate. In contrast with such a case, methods with high precision but lacking accuracy, as at the top right, are extremely unfavorable; there is a real danger that the analyst will be happy with the result, owing to its low standard deviation, and stop looking for sources of error.

In any case the goal is precise and accurate analyses, as at the bottom right, with greater attention to accuracy. For this it is, however, necessary to make use of one's imagination and to take a self-critical approach to one's work.

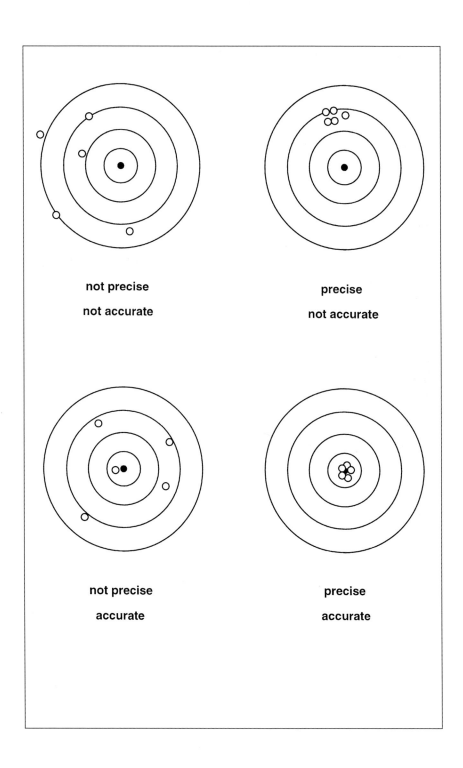

1.10 Standard Deviation

The standard deviation of an analytical result may only be calculated if random deviations ($\rightarrow$ 1.9) occur. The data must be scattered without showing a trend, and systematic effects (due to poor sample preparation, wrong method, a non-adjusted instrument, or even the analyst) must be absent. The true standard deviation σ is only obtainable from an unlimited number of data; what is calculated from a limited set is the "estimate of the standard deviation" s[1]. Nevertheless, s is usually called the standard deviation. s divided by the mean and multiplied by 100 gives the relative standard deviation in percent, also called the coefficient of variation CV. The use of a calculator enables it to be determined with ease.

The significance of the standard deviation is as follows: 68% of the data are within $\pm s$ (only 68%!), 95% are within $\pm 2s$, and 99.7% are within $\pm 3s$. This means that almost one third of the data (32%) deviate from the mean by more than the amount indicated by the value of s.

Example of a data set: 19.92, 19.78, 19.17, 19.03, 19.33, 19.87, 19.83, 19.59, 18.97, 19.31, 19.37, 19.19, 19.25, 19.67, 19.46, 19.27, 19.21, 19.97, 19.76, 19.57. This gives a mean of 19.48 with a standard deviation of ± 0.31 or $\pm 1.6\%$.

Meaning: 68% of the data can be expected between 19.17 and 19.79

32% of the data are presumably smaller than 19.17 or larger than 19.79

95% of the data can be expected between 18.86 and 20.10

5% of the data are presumably smaller than 18.86 or larger than 20.10

[1]

$$s = \pm\sqrt{\frac{d_1^2 + d_2^2 + d_3^2 + \ldots + d_n^2}{n-1}}$$

d: deviation of a data point from the mean
n: number of data

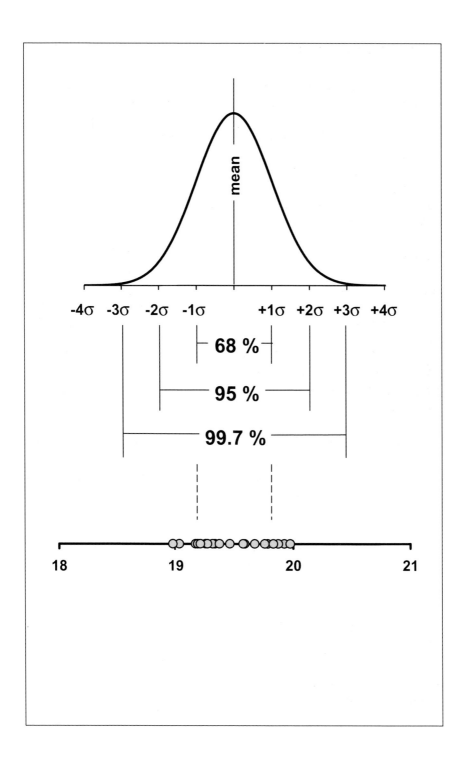

1.11 Uncertainty Propagation

The different kinds of random deviation which occur during an analysis are additive. For a chromatographic determination this means: the weighted mass is slightly too high, the dilution is slightly too low, the injected sample volume is slightly too large, and the peak area as determined by the data system is slightly over-estimated because of an accidental baseline fluctuation. Therefore a value in the upper deviation range is obtained as the final result because it was influenced four times toward a higher number. On the other hand, a result at the lower range is found if all deviations have the opposite sign. Therefore the maximum deviation of a single determination is:

$$\frac{u(R)}{R} = \frac{u(M)}{M} + \frac{u(V)}{V} + \frac{u(I)}{I} + \frac{u(A)}{A}$$

where u = deviation, R = result, M = sample mass, V = volume (dilution steps), I = injection, and A = peak area.

The result will be too high by 4% if all individual deviations are +1%. Examples of 1% deviations are: 101 instead of 100 mg on weighing; 49.5 instead of 50 mL on dilution in the volumetric flask; 202 instead of 200 μL from the pipette; 20.2 instead of 20 μL on injection; 1010 instead of 1000 area counts on integration.

The mean uncertainty of a number of independent analyses is lower because the individual deviations can compensate each other. Therefore variances are added:

$$\frac{u(R)}{R} = \sqrt{\left(\frac{u(M)}{M}\right)^2 + \left(\frac{u(V)}{V}\right)^2 + \left(\frac{u(I)}{I}\right)^2 + \left(\frac{u(A)}{A}\right)^2}$$

For the above-mentioned example a standard deviation of $\pm$ 2% is found ($\rightarrow$ 1.10).

In order to minimize the uncertainty of a result it is advisable to identify the steps or instruments which are most imprecise. If the pipette has an uncertainty of $\pm$ 2% it is necessary to use another brand instead of replacing a balance which works with a precision of 0.1%.

From these considerations it is clear why the less concentrated analytes in a mixture are measured directly and not by difference after a determination of the matrix. If a tablet is composed of 990 mg of excipient and 10 mg of drug, the latter amount is found to be 10 ± 0.4 mg if the uncertainty is $\pm$ 4%. With an indirect determination from the amount of inactive ingredient (excipient) the result would not be satisfactory since this amount would be found to be 990 ± 40 mg.

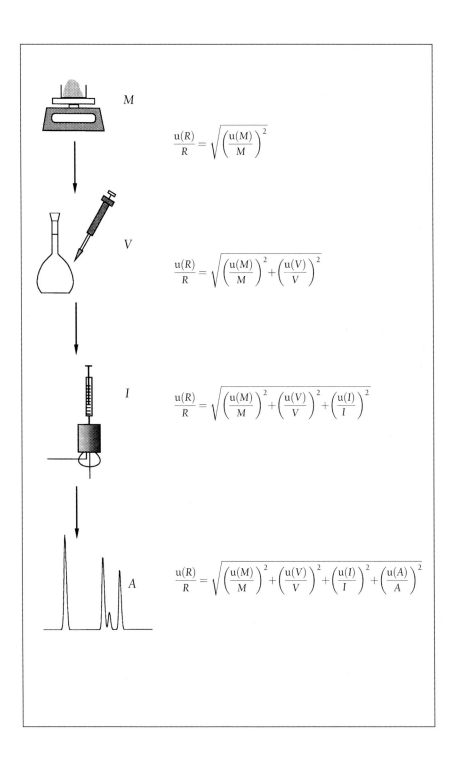

$$\frac{u(R)}{R} = \sqrt{\left(\frac{u(M)}{M}\right)^2}$$

$$\frac{u(R)}{R} = \sqrt{\left(\frac{u(M)}{M}\right)^2 + \left(\frac{u(V)}{V}\right)^2}$$

$$\frac{u(R)}{R} = \sqrt{\left(\frac{u(M)}{M}\right)^2 + \left(\frac{u(V)}{V}\right)^2 + \left(\frac{u(I)}{I}\right)^2}$$

$$\frac{u(R)}{R} = \sqrt{\left(\frac{u(M)}{M}\right)^2 + \left(\frac{u(V)}{V}\right)^2 + \left(\frac{u(I)}{I}\right)^2 + \left(\frac{u(A)}{A}\right)^2}$$

1.12 Reproducibility in Trace Analysis

In 1980 Horwitz, Kamps and Boyer published the results of more than 50 different inter-laboratory tests and found that the relative standard deviation (or coefficient of variation, i.e. the standard deviation divided by the mean) increased strongly with decreasing concentration of the analyte in the sample matrix. The relative standard deviation is here related to the different laboratories participating in a particular test and is, therefore, a measure of inter-laboratory reproducibility of a method (→ 3.11). It does not state anything about the precision of analyses performed at a certain place but represents the scatter of the means determined in different laboratories.

Obviously, the lower the analyte concentration the greater is the danger of obtaining an inaccurate result. This is probably a matter which cannot be altered, but one could try to reduce the extent of the standard deviation. Inter-laboratory tests can be very helpful. Anybody engaged in trace analysis should never cease to think about possible pitfalls and sources of error. The techniques presented in the third part of this book can help us avoid them.

Reference: W. Horwitz, L.R. Kamps and K.W. Boyer: J. Assoc. Off. Anal. Chem. 63 (1980) 1344

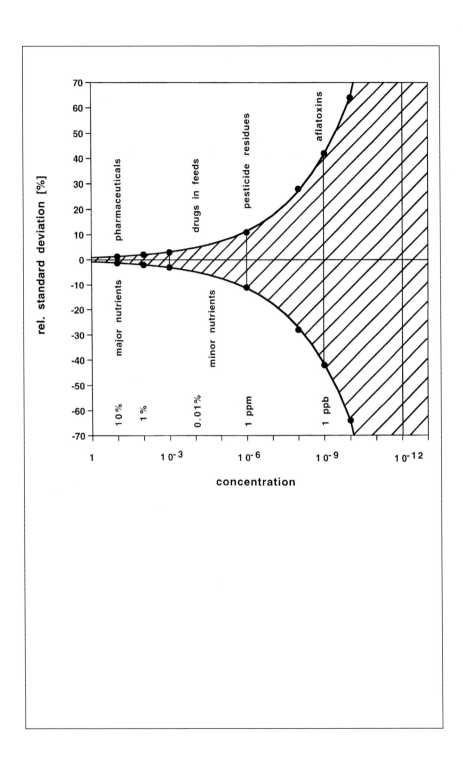

1.13 Ruggedness

A method is rugged if it is insensitive to fluctuations in the separation conditions. For rugged methods, slight changes in temperature, mobile phase composition, pH value and many other parameters, even a deviant sample preparation, have no influence on the analytical result or the chromatogram. Ruggedness needs to be considered if a method is developed for routine use whereas it is of no importance for single analyses.

In most cases it will not be possible to achieve general ruggedness. This would, e.g., also include the type of stationary phase; in the best case it could be possible to perform the analysis on any type of reversed-phase C_{18} without any changes in the method parameters. In practical method development one will try to determine the factors which lead to low ruggedness. Perhaps they can be altered for the better. If not, it is necessary to point them out clearly in the method description ($\rightarrow$ 3.10).

Several examples of inadequate ruggedness can be found in this book. As an illustration of the phenomenon see, e.g., the separation of nicotine and salicylic acid, which is sensitive to the pH of the mobile phase ($\rightarrow$ 2.2), or the analysis of naphthalene and anthracene and the effect of the detection wavelength ($\rightarrow$ 2.34).

Hypothetical example:
The dependence of a separation on the pH and on the concentration of ion-pair reagent in the mobile phase was investigated. As a quality criterion the resolution, R, of the peak pair with poorest separation was chosen and displayed on the z axis. The pH value was varied between 4 and 6.5, and the concentration of ion-pair reagent between 20 and 110 mM. The best resolution was obtained at pH 4.5 and 70 mM; however, this point is an isolated peak which is not at all rugged in any direction. It is advisable to work at pH 5.5–5.75 and 70–80 mM. In this region the resolution is slightly lower but the separation is not influenced by small variations in mobile phase composition.

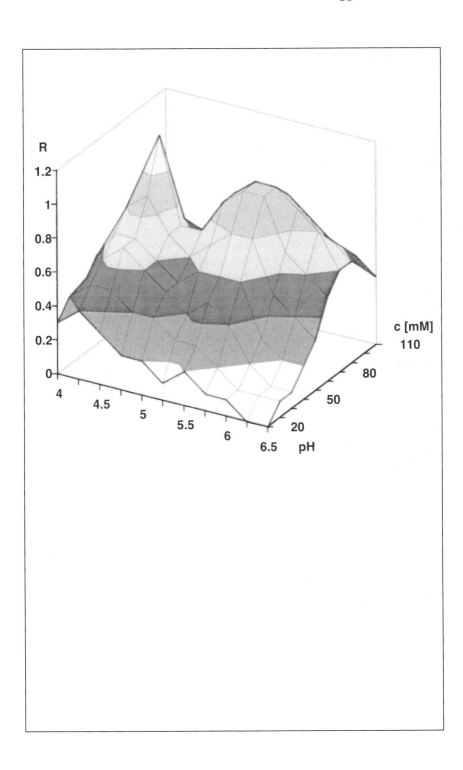

1.14 Calibration Curves

Quantitative analysis demands calibration curves which are linear, run through the origin, and have low scattering of the data points (or a high correlation coefficient, → 2.52). Unfortunately obtaining such a perfect calibration graph is not a matter of course because every method is prone to a number of errors. Different and multiple deviations from the ideal are possible:

Not enough data points: two points only do not result in a calibration curve! The minimum is three points, and five points should be the rule. (This type of error is called sloth and not pitfall.)

High scattering: a repeated determination of the calibration curve from the very beginning will make it clear whether the deviations are of a random or systematic nature. Random effects give other data points, systematic ones will reproduce them at more or less the same spot. In any case the fluctuations should be minimized.

Non-linear curve: if the non-linearity cannot be traced to random effects and if it persists after a thorough checking of the method (→ 3.5), one has no other choice than to use this less favorable curve although attempts should always be made to improve it.

Inadequate spacing of the data points: the calibration points must be spaced evenly over the x axis range. Otherwise the uncertainty is too high in the range without points.

Incomplete calibration curve: this curve does not cover the full span of content which might occur in the samples. It is never permissible to extrapolate a calibration curve to a range which was not investigated.

Calibration curve with proportional-systematic deviation: the slope of the curve is too high (as shown) or too low. This error is not easy to recognize but it affects accuracy! The reason for the deviation can be trivial, e.g. a dilution error, or very unexpected and difficult to explain (→ 2.10).

Calibration curve with constant-systematic deviation: this curve does not run through the origin but is too high (as shown) or too low. The same recommendations are valid as given above for the non-linear curve.

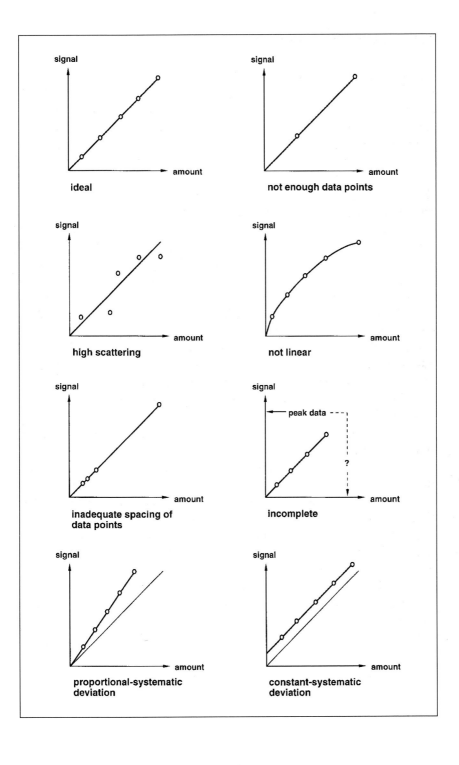

1.15 The HPLC Instrument

The finer the particles of the stationary phase the higher is the number of theoretical plates per unit length ($\rightarrow$ 1.1) and the higher is the optimum flow rate of the mobile phase ($\rightarrow$ 1.5). Therefore HPLC (high performance liquid chromatography) uses stationary phases of 10 μm diameter or less. However, such fine column packings result in high flow resistance, which makes it indispensable to use a pump to transport eluent through the column. A detector enabling observation of the eluted peaks is essential.

Some basic requirements of the individual parts of the HPLC instrument are as follows.

Pump: pulse-free transport of the mobile phase with digital adjustment of the flow rate. The flow must be independent of the back pressure.

Injector: accurate and precise injection of the desired volume must be possible ($\rightarrow$ 2.16). The carry-over between consecutive injections must be negligibly small.

Detector: low noise ($\rightarrow$ 1.17, 2.40, 2.41), wide linear range ($\rightarrow$ 1.16). Neither cell volume nor time constant ($\rightarrow$ 2.50) must contribute markedly to band broadening. The detector can be highly specific (e.g. the electrochemical detector at a defined potential) but also totally non-specific (e.g. the refractive index or the light-scattering detector).

Connecting capillaries: because it is necessary to prevent band broadening, the capillaries from injector to column and from column to detector should be very thin (maximum inner diameter 0.25 mm) and short ($\rightarrow$ 2.20).

Data processing unit: data systems are very convenient and can be used in a quality-assured environment ($\rightarrow$ 3.12, 3.18) if they are adequately designed. They do not, on the other hand, increase the quality of the analytical results above that which can be obtained from a much simpler integrator.

In addition to the items shown here, the HPLC instrument can also be equipped with a gradient system, an autosampler, and a column thermostat.

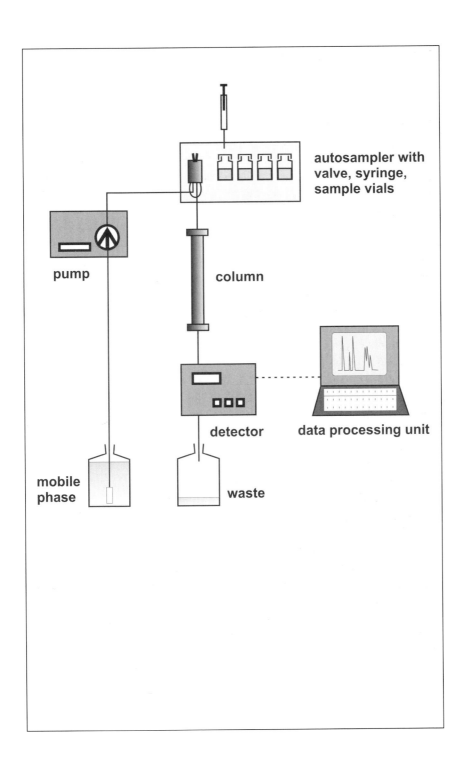

1.16 The Detector Response Curve

As users of a detector, we expect the instrument to yield a signal which is proportional to the amount of sample injected over a wide range of mass or concentration. This so-called linear range is represented by a straight line in a plot of signal (e.g. in millivolts) as a function of mass or concentration. It should stretch over as wide an interval of mass or concentration as possible. Under a certain threshold value the heights of the peaks are very small compared with the noise level, making it impossible to perform an analysis ($\rightarrow$ 1.17, 2.40, 2.41). This is the detection limit, which should be as low as possible but which is directly related to the noise intensity. The detection limit is not identical for qualitative and quantitative analysis. The figure shows a peak approximately at the limit of unambiguous qualitative detection.

Within the linear range the peaks are recorded with their true shape. If more sample is injected than is allowed by the linear range the signal is too flat and not in accordance to its true elution profile. In fact, although there is still a dependence of signal on mass or concentration this relationship is weak and the peaks are not high enough ($\rightarrow$ 2.36). Accurate quantitative analysis is not possible. If the sample amount is even larger, then the dynamic range also is exceeded and the signal is saturated. With a UV detector the photodiode behind the cell does not register any light; with an electrochemical detector the maximum current is running. Under such circumstances the tops of the peaks are cut off horizontally.

A calibration curve must be confined within the linear range ($\rightarrow$ 1.14). In principle it would be possible to expand it to the upper dynamic range but this gives uncertain or even inaccurate results. If necessary the sample needs to be diluted to avoid this. It is totally wrong to extrapolate a calibration curve, starting from a number of data points within the linear range; the upper end of the linear range must be known in order to be sure that samples of high concentration can also be analyzed accurately.

The x axis is for 'mass or concentration' because detectors can be mass or concentration sensitive. UV detectors measure concentration (they are concentration sensitive); if the flow of the mobile phase is stopped at the moment when a peak is passing through the cell the signal remains unchanged. In contrast with such devices coulometric detectors measure a mass flow (they are mass sensitive); if the pump is stopped the signal disappears within a short time.

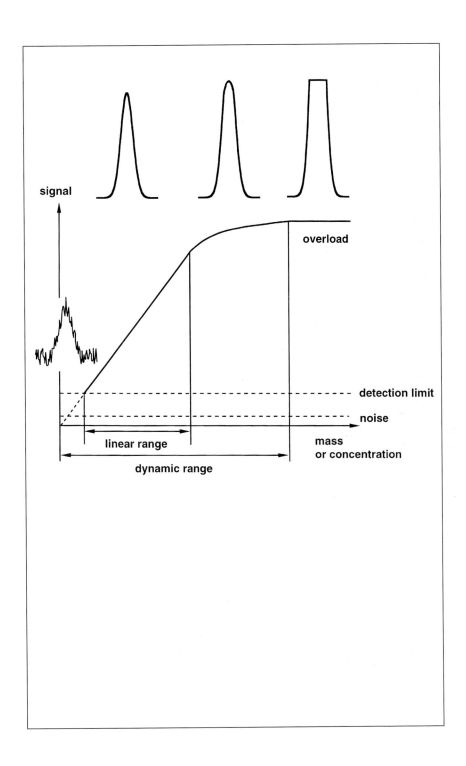

1.17 Noise

Short-term fluctuations of the baseline are called noise. Noise affects quantitative analysis ($\rightarrow$ 2.40, 2.41) and the detection limit in trace analysis. Therefore it is necessary to make every effort to suppress noise: mobile phase degassing (this is especially important with polar eluents and with systems which mix the solvents within the HPLC instrument), the use of a well-maintained, pulse-free pump or of extra pulse-dampeners, the build-up of a small additional pressure drop at the detector output (e.g. by connecting a long capillary of 0.25 mm inner diameter), shielding from interfering electrical fields, and protection of the instrument from draught. Noise can also be suppressed by increasing the time constant of the detector and integrating system ($\rightarrow$ 2.50), but this method impairs the detection of narrow peaks and can only be recommended if both the noise frequency and the width of the narrowest peaks are known. Yet even under the best conditions noise is present which comes from the electronics and which cannot be lower than some threshold level.

The detector response curve ($\rightarrow$ 1.16) disappears at its lower end within the noise and this point determines the detection limit for quantitative and qualitative analysis. The figure presents both limits as they are often defined in the literature: For precise quantitative analysis the signal-to-noise ratio (S/N) should not be lower than 10, as shown by the large peak. This is the limit of quantitation (LOQ). Depending on the analytical demands this value may be too low and it may be better to define the LOQ at S/N = 50 ($\rightarrow$ 2.40). For unambiguous qualitative analysis S/N must not be lower than 3, as for the small peak; this is the limit of detection (LOD). This value also may be too low because in practice the noise is less regular than that shown in the figure, which was calculated by use of a computer generating random numbers within a given span.

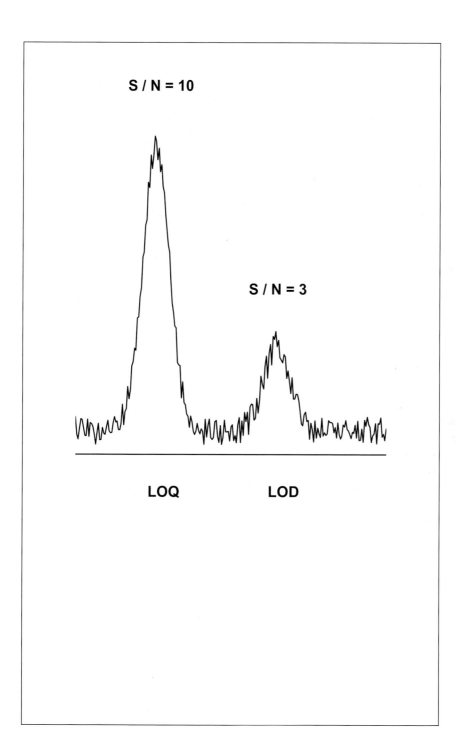

1.18 Causes and Effects Presented as an Ishikawa Diagram

This book is about things that can go wrong when performing HPLC analyses. Every operation or part of the set-up, from the solvent to the presentation on the screen, can be the cause of an erroneous result. The hierarchy of causes which lead to a certain effect can be represented as an Ishikawa diagram (cause-and-effect diagram).

This shows the factors which influence the shape, size, and position (in terms of its retention factor) of a single chromatographic peak. They can be grouped into the effects stemming from injection, separation, detection, and integration. The resulting diagram may be confusing at first sight but there is no question that it could be drawn in even more detail. The complexity of the presentation makes clear that only excellent professional knowledge leads to excellent analyses.

The principle of a quantitative chromatographic analysis consists in the comparison of reference and sample peaks. This approach leads to a certain simplification of the problem as presented in the drawing. If identical volumes of sample and reference solutions are injected, the calibration of the autosampler is irrelevant. Similarly, detection accuracy can now be of minor importance; if two peaks are, e.g., detected at the wrong wavelength, they will both deviate from their "true" values by the same relative amount. If both peaks are of similar size, their shapes and signal-to-noise ratio will be similar, and the integration parameters will be the same anyway. What is important under these circumstances is the repeatability of all steps, including the prerequisite that the mobile phase must not change its properties from one injection to another.

References: K. Ishikawa: Introduction to Quality Control, Kluwer, Dordrecht, 1991, Chapter 7.4.7

ISO 9004-4

V. R. Meyer: J. Chromatogr. Sci. 41 (2003) 439

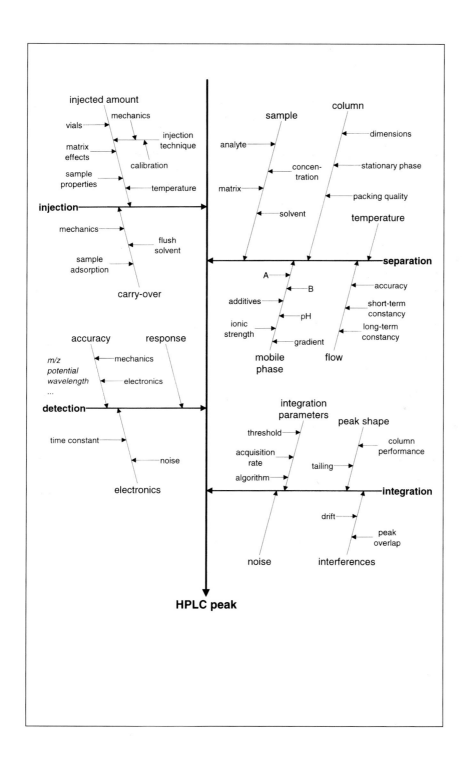

1.19 The Possible and the Impossible

There are only two important questions in chromatography: How many theoretical plates are needed? How long is the analysis time?

The diagrams on the right offer the answers if the analyses are performed at optimum flow rate ($\rightarrow$ 1.5). Under such conditions the necessary pressure is at a minimum. It is, of course, well possible to run a separation at higher speed than the Van Deemter optimum but the plate number will decrease and the pressure will increase (although these deteriorations are low if optimized stationary phases are used).

The diagrams are valid for well-packed columns (reduced plate height $h \approx 2.5$, $\rightarrow$ 1.4), a mobile phase with a viscosity of 1.0 mPa s (water or water with some acetonitrile), a column packing with a porosity of 0.65 (bonded phase), and for analytes with low molecular mass. Continuous lines mark the pressure needed, broken lines the column length, and dotted ones the breakthrough time t_0 ($\rightarrow$ 1.2). In such a graph the magnitude of any two parameters can be chosen ad lib, then the other three are defined.

The upper diagram is a cut-out of the lower one with only the y axis being in logarithmic scale. It becomes obvious that, e.g., 10000 theoretical plates can be realized in various ways, amongst others:

- 1000 bar, 1 µm packing, 3 cm column, and 5 s breakthrough time;
- 50 bar, 5 µm packing, 20 cm column, and approx. 2 min breakthrough time;
- 10 bar, 10 µm packing, 30 cm column, and approx. 10 min breakthrough time.

The lower diagram shows that it is almost impossible to obtain a million theoretical plates by conventional HPLC. As an example, it would be necessary to combine 1000 bar, 10 µm, 3000 cm (a 30 m column), and 30000 s (more than 8 hours). In contrast to these conditions, 2000 theoretical plates at a breakthrough time of 1 s can be realized with a column of not more than 1 cm in length, packed with 1 µm material, and a pressure of approx. 200 bar (if the extra-column volume of the instrument is small, $\rightarrow$ 2.20).

The diagrams are not valid for separations in normal phase (where the circumstances are more favourable due to the low viscosity of the eluent), for monolithic stationary phases (more favourable due to the high permeability of the packing), for ion-exchange chromatography and the like (less favourable due to poor mass transfer), and for separations of macromolecules (less favourable due to low diffusion coefficient).

Reference: I. Halász und G. Görlitz: Angew. Chem. 94 (1982) 50

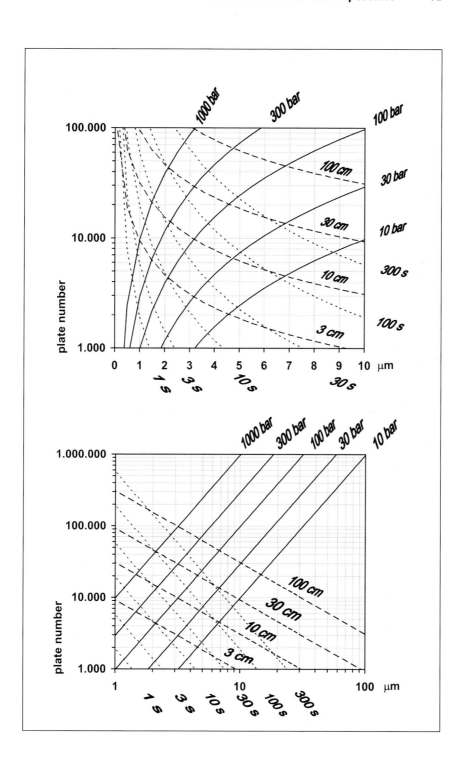

Part II

Pitfalls and Sources of Error

2.1 Mixing of the Mobile Phase

The usual mobile phases used for reversed-phase HPLC sustain a volume contraction when mixed; i.e. the volume of the mixture is smaller than the sum of the volumes of the components. This effect is most pronounced for mixtures of water and methanol (water and acetonitrile or water and tetrahydrofuran are less critical). Different compositions are obtained when the components are measured individually and mixed and when one of the components is poured into a measuring flask and topped with the other. The different mixing ratio then gives rise to mobile phases of different strength, and thus to different retention times.

Here four different methods for the preparation of the mobile phase are compared. They give markedly different retention times.

A: To 400 mL of water methanol was added to give 1 L. Too much methanol!

B: 400 mL of water and 600 mL of methanol were measured individually and mixed. Accurate.

C: In a HPLC system with high pressure gradient (two pumps and mixing chamber) 0.4 mL min^{-1} of water and 0.6 mL min^{-1} of methanol were pumped. The actual flow rate is less than 1 mL min^{-1}; t_0 and t_R increase.

D: To 600 mL of methanol water was added to give 1 L. Too much water!

Chromatographic conditions:

Sample: Explosives dissolved in acetone (octogen, hexogen, tetryl, trinitrotoluene, nitropenta); the first small peak is from acetone

Column: 4.6 mm × 25 cm

Stationary phase: Grom-Sil 80 ODS-7 PH, 4 µm (reversed phase C$_{18}$)

Mobile phase: water/methanol 4 : 6, 1 mL min^{-1}

Detector: UV 220 nm

Reference: V. R. Meyer

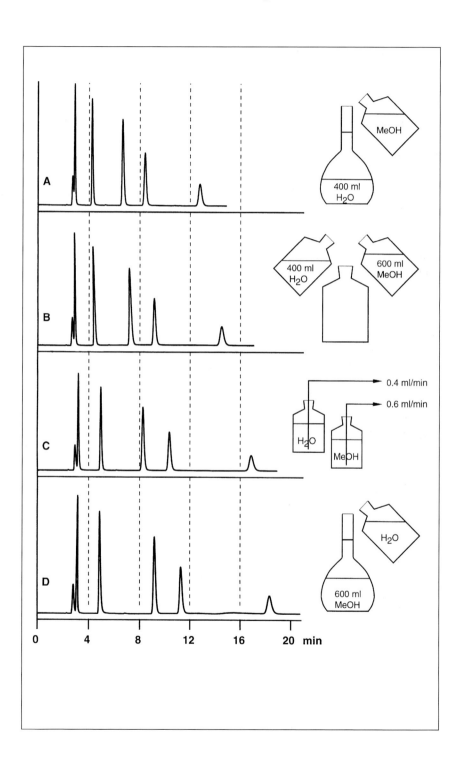

2.2 Mobile Phase pH

Usually it is recommended to control mobile phase pH whenever ionic or ionizable compounds are to be separated. To this end pure water as the polar component of the mobile phase is replaced by buffer. Rugged conditions ($\rightarrow$ 1.13) can be expected if the pH is separated by at least one unit from the pKS of the compound of interest. In principle, although this is not obligatory, non-dissolved species are preferred. For complex mixtures it can be difficult to find the optimum pH.

The separation of nicotine and salicylic acid is not rugged around pH 6; it would be better to work at pH 5 or 7. At pH 5.65 nicotine is eluted in front of salicylic acid, at pH 6.05 the elution order has reversed, and at pH 5.85 the two compounds merge to a single peak. The pKS1 of nicotine is 6.16 (15 °C), the pKS of salicylic acid is 2.96 (25 °C). If 67 mM citrate buffer is used (19.6 g L^{-1} of trisodiumcitrate dihydrate) the addition of 2.2 mL L^{-1} of 25% hydrochloric acid gives pH 6.05, 3 mL L^{-1} gives pH 5.85, and 4 mL L^{-1} gives pH 5.65. Needless to say, such differences in buffer preparation are quite large but errors of this kind can happen when work is performed carelessly.

Note that the nicotine peak has strong tailing if a conventional, not base-deactivated, stationary phase is used.

Chromatographic conditions:

Sample:	nicotine and salicylic acid
Column:	4.0 mm × 25 cm
Stationary phase:	LiChrospher 60 RP-Select B, 5 μm (reversed phase C$_8$)
Mobile phase:	67 mM citrate buffer pH 5.65, 5.85 or 6.05/methanol 6:4, 1.5 mL min^{-1}
Detector:	UV 240 nm

Reference: V. R. Meyer

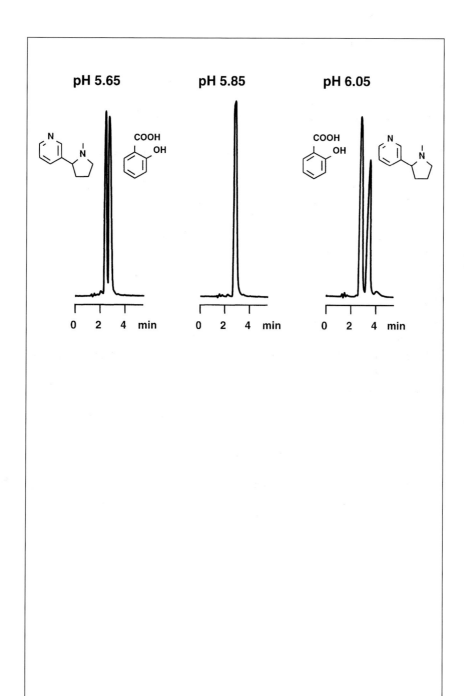

2.3 Adjustment of Mobile Phase *p*H

For the preparation of many aqueous mobile phases not only is the *p*H prescribed but so also is the addition of extra components such as neutral salts (this example), solubility enhancers, ion pair reagents, or organic solvents. At which stage of the preparation should the *p*H be adjusted? It is possible that the separation will be markedly influenced by this decision.

For the separation of α-chymotrypsinogen, cytochrome C, and lysozyme on a strong cation exchanger a citrate buffer of *p*H 4.9 is used; the compounds are eluted by increasing ionic strength, i.e. by a gradient from 0 to 0.5 M sodium chloride. Therefore mobile phase B is also 1 M in neutral salt, and the separation is complete at 50% B.

For the upper chromatogram the *p*H of eluent B was adjusted after addition of the sodium chloride; because of the ionic equilibria of all the species involved the *p*H drops slightly from 4.9 to 4.8 during the gradient. For the lower separation the sodium chloride was added to one part of the buffer stock solution without extra *p*H adjustment. Now cytochrome C and lysozyme are no longer separated and the *p*H drops from 4.9 to 4.5.

The second method cannot be used for this separation and it does not meet scientific standards. (Although not 'scientific', it is not forbidden to lazy people to use this approach but it is obvious that totally new method development would be necessary.) The first method is correct and recommended, yet less convenient. In any case it is necessary to describe mobile phase preparation in detail ($\rightarrow$ 3.11, 3.13).

Chromatographic conditions:

Sample:	0.5 mL solution with α-chymotrypsinogen, cytochrome C and lysozyme
Column:	10 mm × 10 cm
Stationary phase:	Mono S HR, 10 μm (strong cation exchanger)
Mobile phase:	20 mM trisodium citrate *p*H 4.9 with linear gradient from 0 to 0.5 mM sodium chloride (corresponding to 50% B) in 40 min, 4 mL min⁻¹
Detector:	UV 280 nm

Reference: G. Malmquist, Pharmacia Biotech AB, Uppsala, Sweden, poster at the 20th International Symposium on Column Liquid Chromatography, San Francisco, 1996

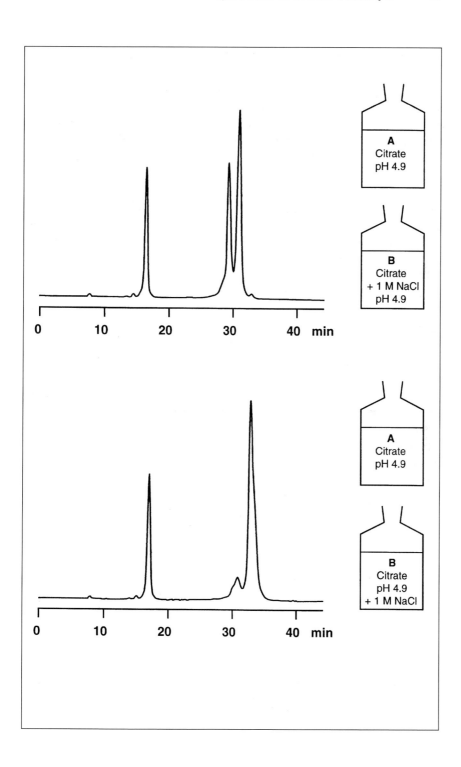

2.4 Inadequate Purity of a Mobile Phase Solvent

All common solvents for liquid chromatography can be bought in so-called HPLC qualities with specified polarity, UV transparency, and purity. In some cases special "gradient grade" solvents are available. They are more expensive than many other qualities but may well be worth the extra money.

One of the problematic solvents is tetrahydrofuran because of its tendency to form peroxides. A common inhibitor is BHT (butyl-hydroxy-toluene, 2,6-di-*tert*-butyl-4-methylphenol). This additive is unwanted for HPLC separations because it appears as a distinct peak in gradient separations.

The quality of commercially available THF, with or without inhibitor, can differ between brands and between batches. This fact may be of minor interest for isocratic separations but with gradients the situation can be unsatisfactory. With poor quality THF, which may nevertheless fulfill the written specifications, ghost peaks will appear during a gradient, even if the amount of THF in the eluent is low.

Chromatographic conditions:

Sample:	none (dummy gradients)
Column:	4.6 mm × 15 cm
Stationary phase:	Zorbax SB-C18, 3.5 µm (reversed phase C_{18})
Mobile phase:	A: water with 0.1% formic acid
	B: methanol with 0.1% formic acid
	C: BHT-free THF with 0.1% formic acid
Gradient:	from 10 to 95% B within 15 min with a constant amount of 5% C, 1.5 mL min^{-1}
Temperature:	40°C
Detector:	UV 265 nm

Reference: S. Williams: J. Chromatogr. A 1052 (2004) 1

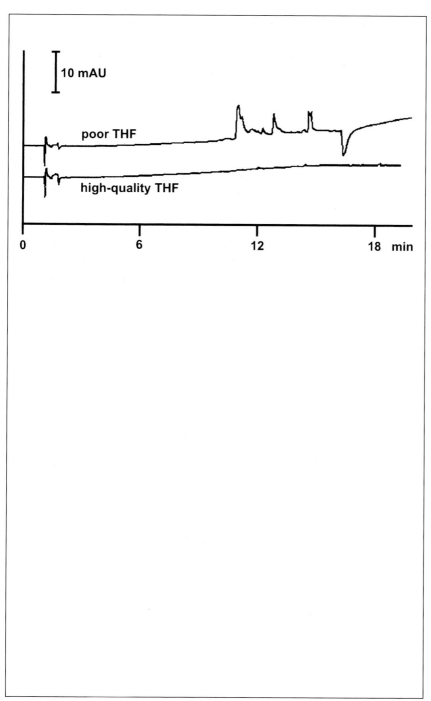

2.5 Inadequate Purity of a Mobile Phase Reagent

All constituents of the mobile phase need to be of such high quality that no extra peaks appear in the chromatogram; with gradient separations, the baseline drift should be negligible. Sometimes this goal is difficult to reach. The usual HPLC solvents are available in adequate purity from many suppliers but it can be critical to find buffer salts, ion pair reagents, and other additives of proper quality. In cases of a detector-active, e.g. UV-absorbing, mobile phase, system peaks can occur ($\rightarrow$ 2.6). Isocratic separations are less prone to interferences than gradient separations; during gradients impurities can be concentrated in the column and later a mobile phase of appropriate composition might elute them as peaks.

The illustration shows dummy gradients (without sample injection) of a separation which was run with ion pair reagents. Three different qualities of sodium dodecyl sulfate did not match purity requirements and gave large extra peaks and baseline disturbances. Fortunately a sodium octane sulfonate could be found which yielded a disturbance-free chromatogram with the exception of a single small peak.

Chromatographic conditions:

Sample:	none (dummy gradients)
Column:	4.6 mm × 15 cm
Stationary phase:	Supelcosil LC 18-DB, 5 µm (reversed phase C_{18})
Mobile phase:	A: 12 mM ion pair reagent as indicated, 3 mL L^{-1} triethylamine, pH 2.5 with phosphoric acid
	B: methanol/acetonitrile 82 : 18
Gradient:	35 % B for 5 min, 35–80 % B from 5 to 20 min, 80 % B from 20 to 23 min, 80–95 % B from 23 to 25 min, then 95 % B; 1 mL min^{-1}
Detector:	UV 280 nm

Reference: J. D. Stafford and B. A. Olsen, Eli Lilly and Company, Lafayette, IN, USA

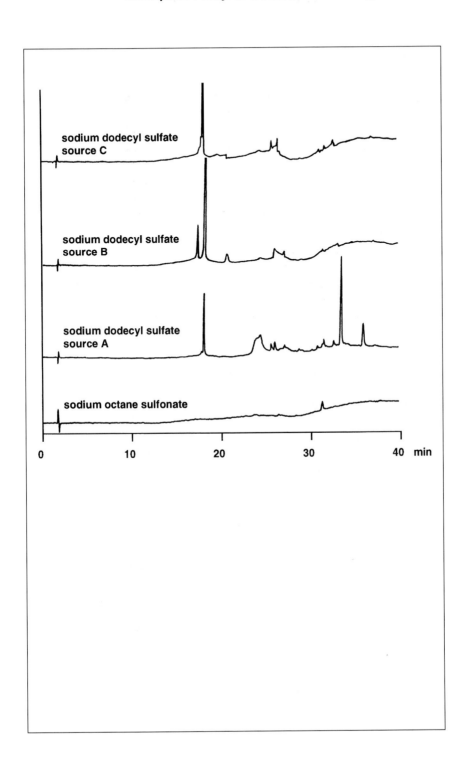

2.6 System Peaks and Quantitative Analysis

With detector-active mobile phases system peaks commonly appear in the chromatogram. "Detector-active" means that the mobile phase itself gives rise to a detector signal; with a UV detector, the eluent absorbs at the chosen wavelength. System peaks are additional peaks in the chromatogram which appear at a particular retention time (in contrast to ghost peaks, which have random retention times). They do not belong to a sample compound but are characteristic of a mobile phase component.

System peaks are unwanted because any additional peaks increase the difficulty of the separation. They are, moreover, the reason for another unwanted effect, as can be seen in the figure. The peak areas of the regular peaks do not only depend on the mass of sample injected but also on their position relative to the system peaks.

The separation of racemic bupivacaine on a chiral stationary phase was detected at 215 nm. At this low wavelength the mobile phase used has some UV absorption, coming maybe from isopropanol or from buffer reagents of inadequate quality. Two system peaks are observed, one with a negative signal. Because a racemate was injected the peak areas of both enantiomers should be identical but obviously this is not so. If a proper calibration curve has previously been established it will be possible to obtain the accurate analytical result; otherwise the ratio of enantiomers calculated by the integrator will be wrong.

Chromatographic conditions:

Sample:	racemic bupivacaine
Stationary phase:	EnantioPac, 10 µm
	(α_1-acid glycoprotein, chiral phase)
Mobile phase:	phosphate buffer pH 7.2/2-propanol 92 : 8
Detector:	UV 215 nm

Reference: G. Schill and J. Crommen: TrAC 6 (1987) 111

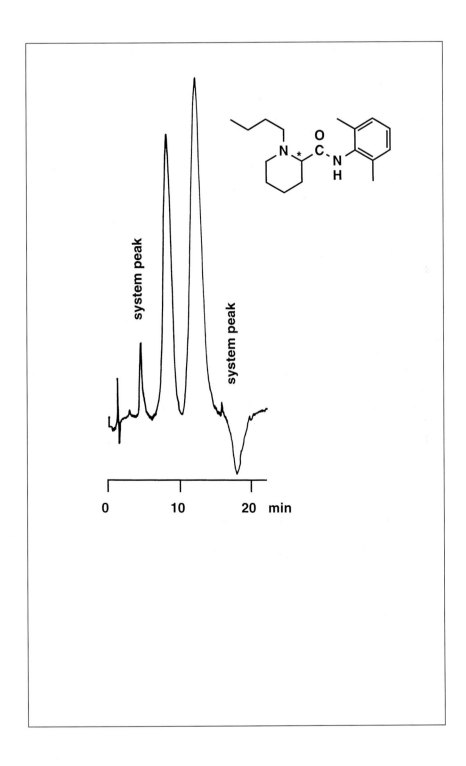

2.7 Sample Preparation with Solid Phase Extraction

The steps involved in sample preparation can be the source of numerous errors. The problem shown here is only a single example of how a chromatogram can be influenced drastically by a deviation from the recommended procedure.

For the isolation of tricyclic antidepressants a solid phase extraction step was necessary. The cartridge was wetted with 1 mL methanol followed by 1 mL water. After application of 1 mL sample solution and a washing step with 1 mL 5 % methanol in water the antidepressants were eluted with 1 mL methanol. The solvent was evaporated and the residue dissolved in 200 µL phosphate buffer/methanol 80 : 20. This correct procedure gives the upper chromatogram.

If, however, a solution of 2 % phosphoric acid in methanol was used for elution from the cartridge the acid was concentrated during evaporation, and therefore the drugs became protonated and the final sample solution was ca. 10 % in phosphoric acid. Thus the retention times of the tricyclic antidepressants are shorter and the peaks are distorted as seen in the lower chromatogram.

Chromatographic conditions:

Sample:	20 µL solution with tricyclic antidepressants (nordoxepin, nortriptyline, doxepin, amitriptyline)
Column:	3.9 mm × 15 cm
Stationary phase:	Symmetry C18, 5 µm (reversed phase C_{18})
Mobile phase:	20 mM phosphate buffer pH 7/methanol 32 : 68, 1 mL min^{-1}
Temperature:	35 °C
Detector:	UV 254 nm

Reference: Y. F. Cheng, Waters, Milford, MA, USA

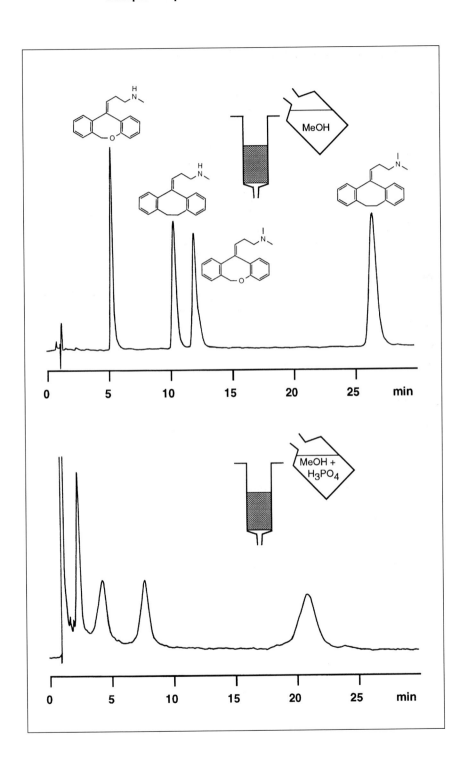

2.8 Poor Choice of Sample Solvent: Peak Distortion

Samples for HPLC analysis usually need to be dissolved in a suitable solvent, either to furnish a solution for analysis or to obtain a convenient dilution. The best general recommendation is to dissolve the sample in the mobile phase itself; this protects against unwanted phenomena such as precipitation within the column, excessive band broadening, or unusual peak shapes. With gradient separations the sample solvent should correspond to the mobile phase composition at the time of injection. One exception, however, is usually allowed: the solvent may be weaker than the mobile phase. For reversed-phase chromatography this means more water, for normal-phase chromatography, more hexane, than is present in the mobile phase. This trick can even be used to concentrate the samples at the column entrance, thus making the peaks narrower.

The example shows the analysis of phenylalanine, which needed a very weak mobile phase containing only 8% acetonitrile. Dissolving the sample in pure water resulted in a satisfactory peak shape. When the acetonitrile content of the sample solution was higher than that of the mobile phase, peak broadening and even very strange distortion of the phenylalanine peak were observed. It is possible that the effects were intensified by the very low detection wavelength of 210 nm.

Chromatographic conditions:

Sample:	phenylalanine, 20 µL with 0.9 µg of Phe dissolved in various mixtures of water and acetonitrile
Column:	4.6 mm × 25 cm
Stationary phase:	ODS, 5 µm (reversed phase C_{18})
Mobile phase:	5 mM phosphate buffer pH 3.5/acetonitrile 92 : 8, 1 mL min^{-1}
Detector:	UV 210 nm

Reference: N. E. Hoffman, S. L. Pan and A. M. Rustum: J. Chromatogr. 465 (1989) 189

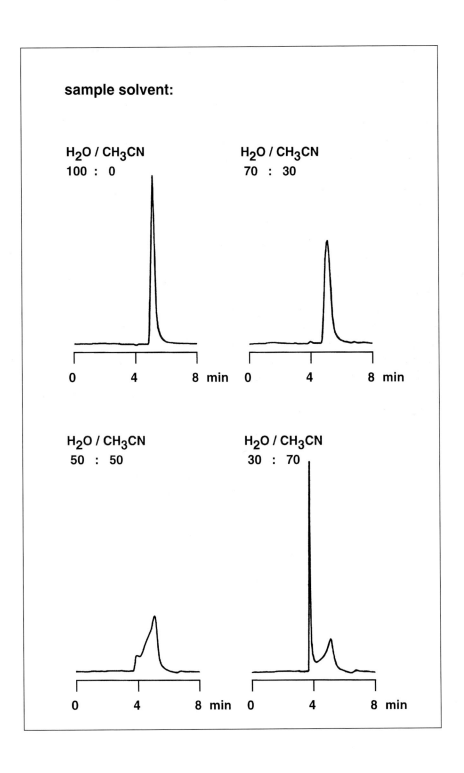

2.9 Poor Choice of Sample Solvent: Tailing

Although an inappropriate sample solvent does not necessarily result in severe peak distortion ($\rightarrow$ 2.8), more subtle impairment of peak shape is also unwanted. During analysis of fat-soluble vitamins, tailing was observed at the very bottom of the peaks if the sample was dissolved in methanol instead of mobile phase. This is surprising because the eluent was almost pure organic solvent with a water content of only ca. 1%.

Tailing is always unwanted because it reduces peak height, impedes the detection of trace components, reduces the resolution of adjacent peaks ($\rightarrow$ 2.48) and, as the worst effect, makes peak integration more difficult. The integrator, with its programmed algorithms, has more difficulty defining the end of the peak as tailing increases; accurate integration can be impossible. With a data system the peak start and end can be manipulated subsequently but, especially for the circled peak in the upper chromatogram, it is difficult for man or computer to define the peak end.

Chromatographic conditions:

Sample:	uracil (as breakthrough marker), vitamin A, vitamin D_3 and vitamin E, dissolved in methanol or in mobile phase
Column:	3.9 mm $\times$ 15 cm and precolumn 3.9 mm $\times$ 2 cm
Stationary phase:	Symmetry C_8, 5 μm (reversed phase C_{18})
Mobile phase:	water/methanol/acetonitrile 1 : 25 : 25, 1 mL min^{-1}
Detector:	UV 280 nm

Reference: U. D. Neue, D. J. Phillips, T. H. Walter, M. Capparella,
B. Alden and R. P. Fisk: LC GC Int. 8 (1995) 26

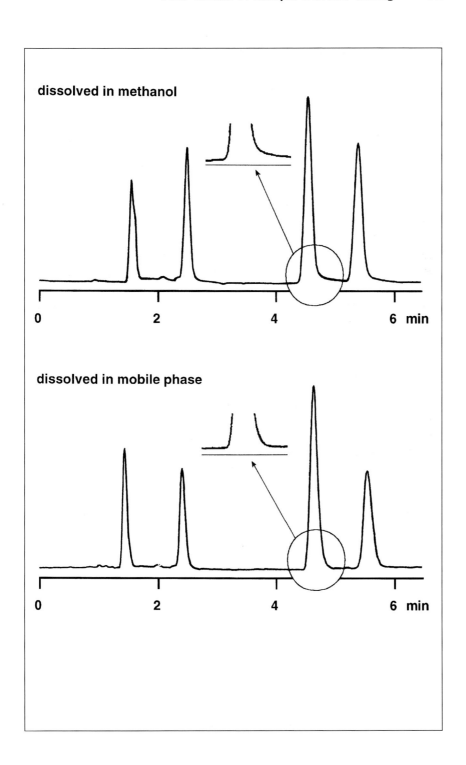

2.10 Sample Solvent and Calibration Curve

Sometimes in analytical HPLC one experiences unexpected effects which are difficult to explain. In the example presented here fosinopril sodium was separated on silica with an aqueous mobile phase; this is an unusual but nevertheless valid phase system. The detection wavelength was, moreover, very low. The analysts found that the peak areas were very strongly dependent on the particular sample solvent although not more than 20 µL were injected. They could not explain this effect. As long as the same sample solvent is used for calibration and analysis no problems will arise with this method. If for some reason, however, the solvent is changed for the analyses the results will be extremely inaccurate (although precise). It is, therefore, indispensable to prescribe the sample solvent for calibration curve and analysis and to mention the problem in the method description (→ 3.10, 3.11).

As long as the circumstances have not been studied it is never possible to exclude such phenomena, not even with conventional phase systems. In this example the possible analytical errors are serious; with another method they could be much smaller but not negligible.

Chromatographic conditions:

Sample:	fosinopril sodium, 20 µL solution in ethanol, methanol or water
Column:	3.9 mm × 15 cm
Stationary phase:	Resolve Silica, 5 µm (silica)
Mobile phase:	acetonitrile/water/orthophosphoric acid 4000 : 15 : 2, 1 mL min^{-1}
Detector:	UV 205 nm

Reference: J. Kirschbaum, J. Noroski, A. Cosey, D. Mayo and
J. Adamovics: J. Chromatogr. 507 (1990) 165

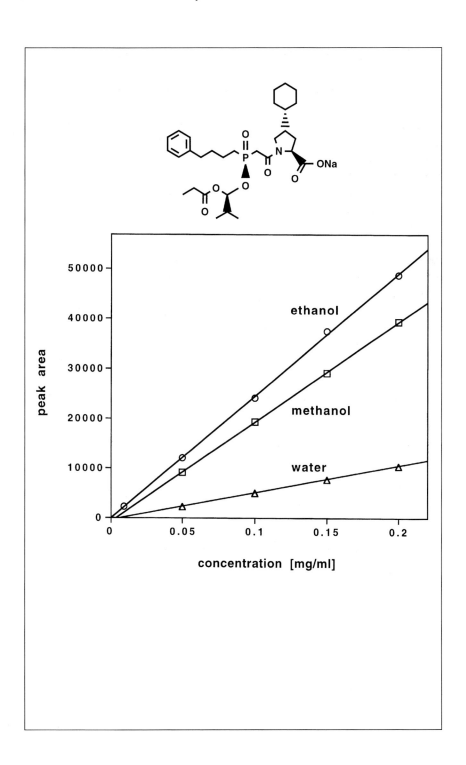

2.11 Impurities in the Sample

The signal of a sample component in the UV detector is proportional to its molar absorptivity ε and its concentration. The absorptivity is a function of the wavelength, as every UV spectrum shows, and, of course, depends on the particular molecule ($\rightarrow$ 2.33). It is, therefore, never permissible to postulate a certain mass ratio from a given peak-size ratio.

An extreme example is the pair chlorostilbene oxide and chlorobenzophenone if detection is at 254 nm. The commercially available *trans*-4-chlorostilbene oxide (first peak) has a purity of 98% and contains some 4-chlorobenzophenone as an impurity (plus several other trace components). As the UV spectra show, chlorostilbene oxide has much lower absorptivity at 254 nm (dotted line) than does chlorobenzophenone. This leads to almost identical peak heights of main component and impurity at the chosen wavelength. When analysis was performed on a chiral stationary phase the two peaks were taken erroneously as the enantiomers of chlorostilbene oxide.

The authors of the paper cited below list two other pairs of compounds as similar examples: *trans*-stilbene oxide/benzophenone and α-methoxy-α-(trifluoromethyl)phenylacetonitrile/2,2,2-trifluoroacetophenone.

Chromatographic conditions:

Sample:	*trans*-4-chlorostilbene oxide (Aldrich, 98%), containing 4-chlorobenzophenone as impurity, dissolved in mobile phase
Column:	3.2 mm × 25 cm
Stationary phase:	LiChrosorb SI 60, 5 μm (silica)
Mobile phase:	hexane/tetrahydrofuran 99 : 1, 1 mL min^{-1}
Detector:	UV 254 nm
UV spectra:	40 μM solutions in mobile phase

Reference: V. R. Meyer, after L. Oliveros and C. Minguillón:
J. Chromatogr. A 653 (1993) 144

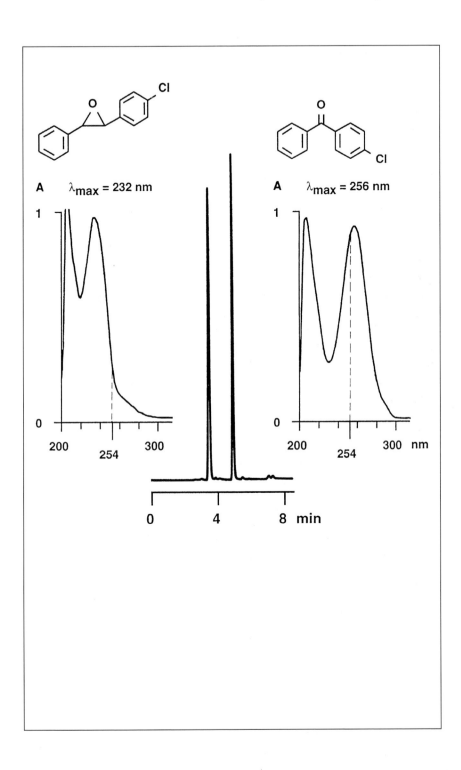

2.12 Formation of a By–Product in the Sample Solution

An irksome analytical error is the formation of a by-product during sample preparation or in the solvent which is used for injection. By such a reaction the content of the analyte decreases and an additional peak might appear in the chromatogram (although this is not inevitable because it is possible that the by-products will not be eluted under the given conditions). The reaction can be time-dependent; then the error is larger the longer the time necessary for sample preparation or the longer the sample will be stored between preparation and injection. Sample storage at low temperature can be advantageous.

Primary amines can react with aldehydes and ketones to give Schiff bases. This was observed during the analysis of bendroflumethiazide when the sample was dissolved in methanol containing a trace of formaldehyde. The concentration of formaldehyde was not higher than $3 \, \mu L \, L^{-1}$ but this small amount was enough to produce the azomethine which appeared then as a small peak in the chromatogram. The area and height of the bendroflumethiazide peak decreased accordingly. With formaldehyde-free methanol the problem was not observed.

Chromatographic conditions:

Sample:	bendroflumethiazide, dissolved in methanol with or without formaldehyde
Stationary phase:	phenyl
Mobile phase:	water with 0.1 M NaCl and 25 mM sodium acetate/methanol 6 : 4
Detector:	UV 270 nm

Reference: J. Kirschbaum, S. Perlman and R.B. Poet: J. Chromatogr. Sci. 20 (1982) 336

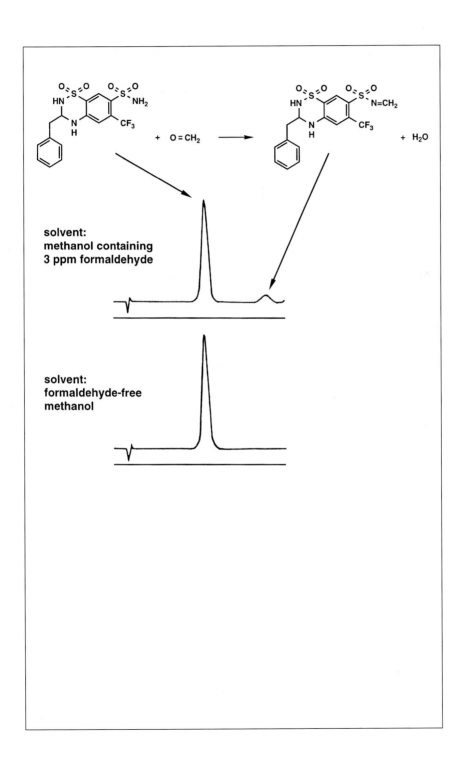

2.13 Decomposition by the Sample Vial

The driving force of analyte decomposition can be an unexpected one, in this case the type of glass of the vial.

Carnosic acid, a main component of rosemary extract, is known to have weak stability because it is prone to oxidation, especially when dissolved in methanol. In order to protect the samples from light, they were stored in amber autosampler vials. The observed decomposition, i.e. the decrease of the peak area over a time period of 30 h, was higher than expected. In fact, the decomposition was markedly lower in clear glass even when no extra light protection was used.

Amber glass has significant contents of iron and titanium: $0.7-1\%$ Fe_2O_3 and $3-5\%$ TiO_2. In contrast to these high amounts, clear glass (for vials) has only 0.04% Fe_2O_3. Obviously, the transition metal cations act as catalysts for the oxidation of carnosic acid.

Chromatographic conditions:

Sample for stability test:
                carnosic acid of 99% purity, dissolved in methanol, 293 µg mL^{-1}

Sample for chromatogram:
                carnosol and carnosic acid, dissolved in dimethyl sulfoxide

Column: 3.0 mm × 15 cm

Stationary phase: Zorbax SB-C$_{18}$, 3.5 µm (reversed phase C$_{18}$)

Mobile phase: water/acetonitrile/trifluoroacetic acid 40 : 60 : 0.15, 0.42 mL min^{-1}

Vial temperature: ambient

Column temperature:
                45°C

Detector: UV 230 nm

Reference: M. A. Thorsen and K. S. Hildebrandt: J. Chromatogr. A 995 (2003) 119

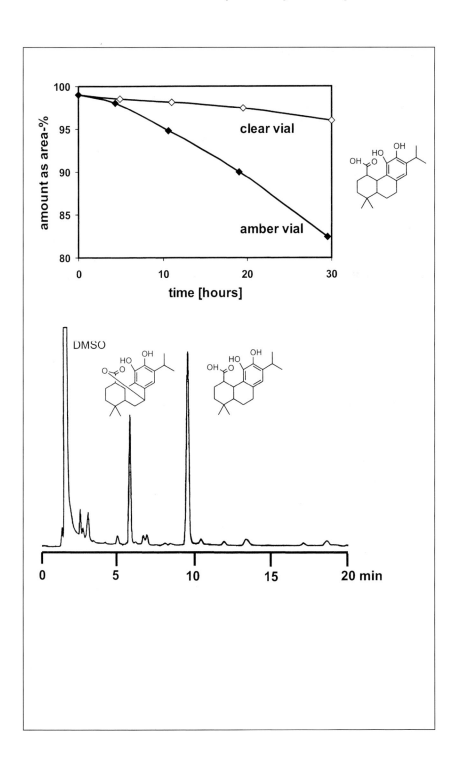

2.14 Artifact Peaks from the Vial Septum

Sample vials for autosampler injection need to be closed with a septum cap. The septum allows a tight closure even after puncture with the syringe needle. If the septum elastomer is not properly chosen with regard to the sample solution it is possible that artifact peaks will appear in the chromatogram.

For the analysis of impurities in a new drug substance the vials were closed with teflon-lined rubber septa. The first injection showed a large peak of the active compound plus the minor peaks of its impurities from synthesis (upper chromatogram). However, with the second injection from the same vial some additional small peaks showed up in the chromatogram; for investigation purposes the vial was shaken vigorously before the second injection was made (lower chromatogram). If teflon-lined silicone septa were used no extra peaks were found even if the liner was punctured and the vial was shaken. With the first set-up several compounds could be extracted from the rubber once the teflon protection layer was damaged, whereas the silicone material was inert.

Chromatographic conditions:

Sample:	50 µl of drug solution (active compound with impurities), approx. 2 mg mL^{-1} in mobile phase
Vial septum:	teflon-lined rubber
Column:	4.6 mm × 15 cm
Stationary phase:	Symmetry C_{18}, 5 µm (reversed phase C_{18})
Mobile phase:	10 mM phosphate buffer pH 6.0/acetonitrile 60 : 40, 1 mL min^{-1}
Temperature:	40°C
Detector:	UV 220 nm

Reference: G. R. Strasser and I. Váradi: J. Chromatogr. A 869 (2000) 85

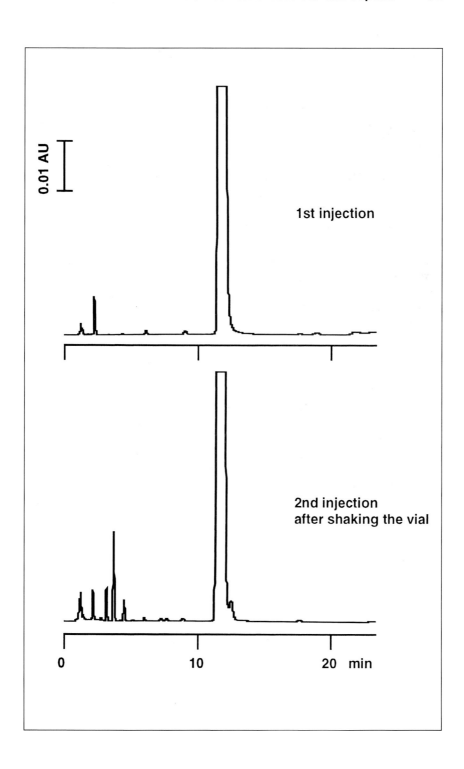

2.15 Formation of an Associate in the Sample Solution

A rare source of error is the formation of an associate from two sample compounds which then appears as *one* peak in the chromatogram. The associate behaves like a molecule and cannot be separated with the given phase system.

As an example, methylprednisolone (top) and tetrahydrocortisone (middle) are such a pair because they can form a total of four intermolecular hydrogen bonds in the preferred configuration. The interaction energy is rather high, namely -137 kJ mol^{-1}. For the individual steroids the retention factors are 2.36 and 2.26, respectively, and the peaks are broad. If the mixture is injected the retention factor of the associate drops to 2.02; obviously the interaction with the stationary phase is now more favorable and the peak is narrow. An analogous effect is observed with cortisone (with two keto groups in positions 3 and 11 of the steroid skeleton) and tetrahydrocortisol (with two hydroxyl groups).

Chromatographic conditions:

Sample:	methylprednisolone and tetrahydrocortisone dissolved in acetonitrile
Column:	4.6 mm × 47 cm
Stationary phase:	Spheri-5 RP-18, 5 μm (reversed phase C$_{18}$)
Mobile phase:	water/acetonitrile 65 : 35, 0.5 mL min^{-1}
Detector:	UV 240 nm

Reference: P.H. Lukulay and V.L. McGuffin: J. Liquid Chromatogr. 19 (1996) 2039

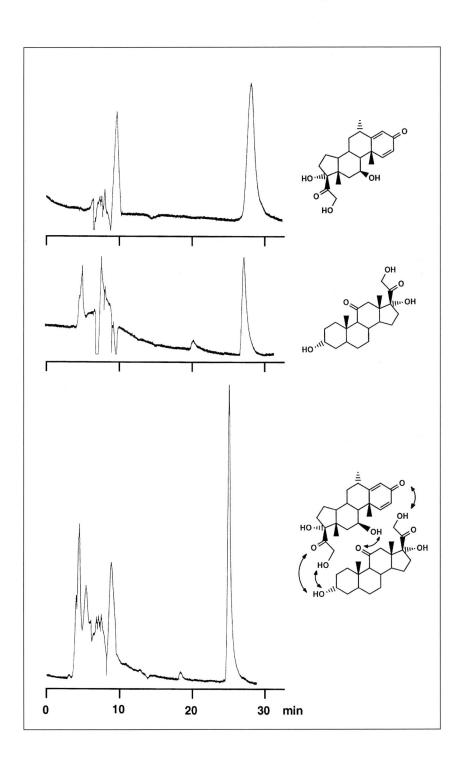

2.16 Precision and Accuracy with Loop Injection

Injection valves are equipped with a loop which is filled with a syringe either manually or by means of an automated machine. With a subsequent turn of the valve the liquid in the loop is transferred to the column by the flowing mobile phase. The loop is, therefore, full of eluent before the next sample is applied.

When the loop is fed with sample an adverse effect takes place: the sample solution is partially mixed with the solvent which it must replace. As a consequence less than, e.g., 20 µL of sample are confined in a 20 µL loop if 20 µL had been applied because a small amount penetrates into the overfill capillary and some old solvent still remains. For this reason the manufacturers of injection valves recommend, for quantitative analyses, either filling the loop with not more than 50% of its volume or, if enough sample is available, injecting fivefold its volume.

In the experiment with a 20 µL loop it was found that accuracy is only achieved with approx. 100 µL of sample and that 200 µL or more is better (upper graph). The accuracy is especially poor if the exact loop volume (20 µL) is injected. With 40 µL and more the precision is better than 0.5%; with smaller volumes it depends on the capability of reproducible injection (lower graph).

The poor accuracy (although not the precision) is not of importance if the analysis is performed with an external standard and with constant volume (!). If an internal standard is used neither injection accuracy nor precision has any an influence on the result.

Chromatographic conditions:

Sample:	thiourea, 50 µg mL^{-1} in water
Injection:	manual injection, means of 5 determinations, 20 µL loop, Rheodyne 7125 valve
Column:	3.2 mm × 25 cm
Stationary phase:	Spherisorb ODS, 5 µm (reversed phase C$_{18}$)
Mobile phase:	water, 1 mL min^{-1}
Detector:	UV 238 nm

Reference: V. R. Meyer, see also J. W. Dolan: LC Magazine 3 (1985) 1050

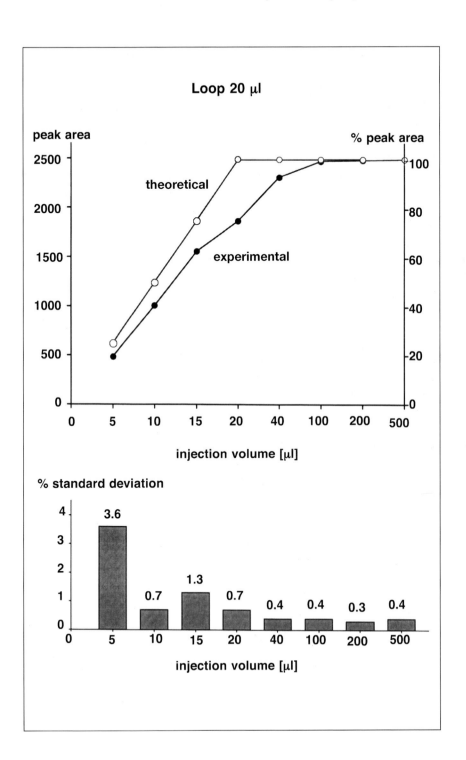

2.17 Injection Technique

With manual injection the question arises whether the syringe should be kept within the injector or removed during switching. As can be seen from the data in the figure it is not possible to achieve high precision if the loop is filled partially ($\rightarrow$ 2.16) and the syringe is removed.

When the syringe is pulled out of the injector one cannot avoid withdrawing some sample solution back into the needle channel. With partial loop filling a fraction of the carefully injected sample is lost; the volume of this loss is unknown and not defined, and therefore the analytical precision (here given as relative standard deviation) decreases (top left). This sample imprecision would also be observed with 100% loop fill. Only with adequate overfill, as e.g. with the fivefold volume shown here, can enough sample solution flow back from the outlet capillary into the loop (bottom). Under these circumstances it makes no difference whether the syringe remains in the loop or not (the indicated differences in precision are not significant).

It is, nevertheless, a recommended habit to leave the syringe in the injector whenever quantitative analysis is performed because it isolates the loop from the exterior if needle and seal fit correctly. Remember: if the end of the outlet capillary is not positioned at the same height as the needle seal there is a danger that sample will be lost by a siphon drain when the syringe is removed before the valve is turned.

Chromatographic conditions:

Sample:	nitrobenzene dissolved in mobile phase, $k = 4$
Injection:	manually, means from 10 injections, 10 µL (top) or 100 µL (bottom) into a 20 µL loop, Rheodyne 7125 valve
Column:	4.6 mm $\times$ 7.5 cm
Stationary phase:	Zorbax SB C-18, 3.5 µm (reversed phase C_{18})
Mobile phase:	water/acetonitrile 6 : 4, 2 mL min^{-1}
Detector:	UV 280 nm

Reference: V. R. Meyer

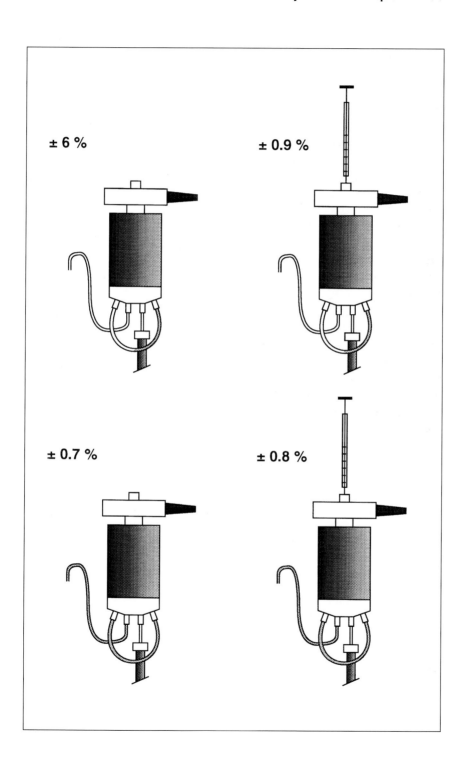

2.18 Injection of Air

If several microlitres of air are injected, a small, retained peak with severe tailing will appear in the chromatogram if detection is performed at low UV wavelength (not to be confused with signals generated by air bubbles from a poorly degassed eluent). If, because the analyte concentration is high, the chromatogram can be recorded at low attenuation, one will not be aware of this peak. With low concentrations and trace analysis it can be disturbing.

Although nobody injects air intentionally this can happen with careless manual injection, if the autosampler needle is not properly adjusted to the vials in use, if the flush solvent reservoir is empty, or if the vials are not filled with enough sample.

Chromatographic conditions:

Sample:	solution of octogen in water and a little acetone
Column:	2 mm × 15 cm
Stationary phase:	YMC 120 ODS-AQ, 3 μm (reversed phase C_{18})
Mobile phase:	water/acetonitrile, 67 : 33, 0.3 mL min^{-1}
Detector:	UV 210 nm

The small peak in the lower chromatogram which appears at the appropriate retention time is probably caused by dissolved air in the sample.

Reference: V. R. Meyer, see also J. W. Dolan, D. H. Marchand and S. A. Cahill: LC GC Int. 10 (1997) 274 (with printing errors: mL instead of μL)

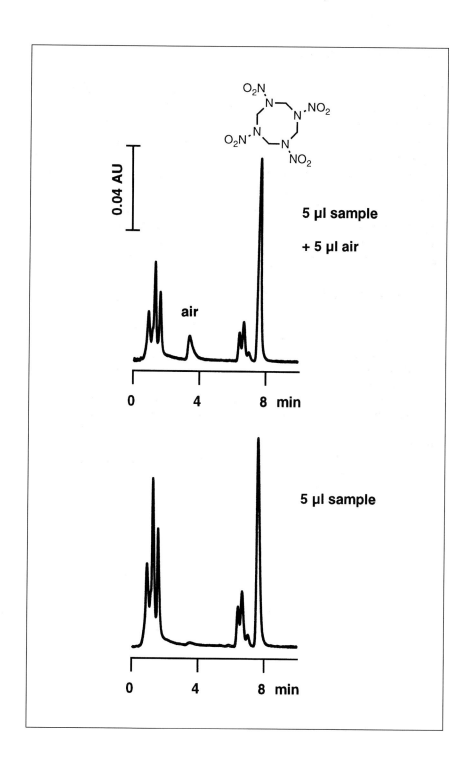

2.19 Sample Adsorption in the Loop

If the analyte is only poorly soluble in the sample solvent it can become adsorbed on the inner wall of the injector loop (and of other capillaries); basic compounds are possibly more prone to such effects than others. If the mobile phase is different and is a better solvent it might be expected that the adsorbed compounds will be desorbed during the injection process although an unpleasant surprise cannot be excluded.

It is possible that such reversible adsorption would not affect the analysis and would, in fact, not even be recognized. Problems will, however, arise if the recommended technique of loop overfill ($\rightarrow$ 2.16) is used. If terfenadine is dissoved in water (a poor solvent for this compound), the peak area increases with increasing sample volume. This means that the advantage of loop overfill, i.e. excellent reproducibility, is lost; now it is necessary to inject the standard and sample solutions with high volume precision. As usual, the recommended procedure is to dissolve the sample in the mobile phase – then the peak area is independent of sample volume, even with a 15-fold overfill. For this analysis the same is even true for methanol as sample solvent; methanol is a stronger eluent than the mobile phase (although it should be remembered that stronger solvents are often unsuitable and give rise to other problems ($\rightarrow$ 2.8, 2.9)).

The authors of the study found analogous results for astemizole, bromhexine, and ergotamine, all molecules with amine nitrogen; with phenylephrine, caffeine, betamethasone, ambroxol, and enapril, on the other hand, no problems were observed.

Chromatographic conditions:

Sample:	terfenadine, 5 µg mL^{-1} in water, mobile phase or methanol
Injection:	60, 150 or 300 µL in a 20 µL loop (stainless steel)
Column:	4.6 mm $\times$ 15 cm
Stationary phase:	MicroPak MCH-5, 5 µm (reversed phase C$_{18}$)
Mobile phase:	0.03 M phosphate buffer pH 3.0/acetonitrile 7 : 3, 1.7 mL min^{-1}
Detector:	UV 230 nm

Reference: G. C. Fernández Otero, S. E. Lucangioli and C. N. Carducci: J. Chromatogr. A 654 (1993) 87

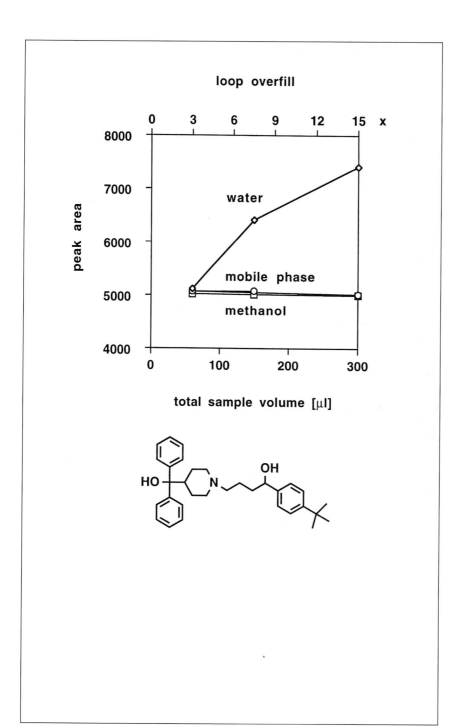

2.20 Extra-Column Volumes

The separation performance of the column is affected by the extra-column volumes (also called dead volumes) of the instrument; these should, therefore, be kept as low as possible. All elements of the HPLC instrument between injector and detector are part of the extra-column volume, i.e. several fittings, two capillaries (or even more), the column frits, and the detector cell. As the user of an HPLC system one can usually only modify the capillaries because the design of the column end fittings is fixed. Sometimes, however, it is worthwhile inspecting the detector also because many instruments are equipped with long, winding internal capillaries to improve heat exchange. Then a new and short capillary can be built in, a measure which will reduce this part of the extra-column volume; it is necessary to check whether noise increases markedly. If so it is necessary to protect the instrument from short-term temperature fluctuations (e.g. draught) or to thermostat it.

The capillaries between injector, column, and detector should have a maximum inner diameter of 0.25 mm. The inner diameter d_c is far more important than the length! It influences the separation performance by the fourth power, i.e. by d_c^4. It is not really detrimental to use a capillary of double length if this is necessary because of the use of a short column. On the other hand, replacement of 0.25 mm capillary by one of 0.5 mm inner diameter can affect the chromatogram as the example shows, even when a column of moderate performance is used. Peaks with small capacity factors thereby undergo more broadening than later eluted ones.

Chromatographic conditions:

Sample:	8 µL of solution of methyl, ethyl, propyl, and butyl parabene
Column:	4.0 mm × 10 cm
Stationary phase:	Nucleosil 5 C18, 5 µm (reversed phase C_{18})
Mobile phase:	water/methanol 1 : 3, 1.2 mL min^{-1}
Detector:	UV 254 nm
Resolution:	R_{12} (t_R = 1.02 min and 1.13 min) = 0.85 and 0.7, respectively (N_1 = 1400 and 1000, respectively) R_{34} (t_R = 1.32 min and 1.63 min) = 1.9 and 1.7, respectively

Reference: V. R. Meyer

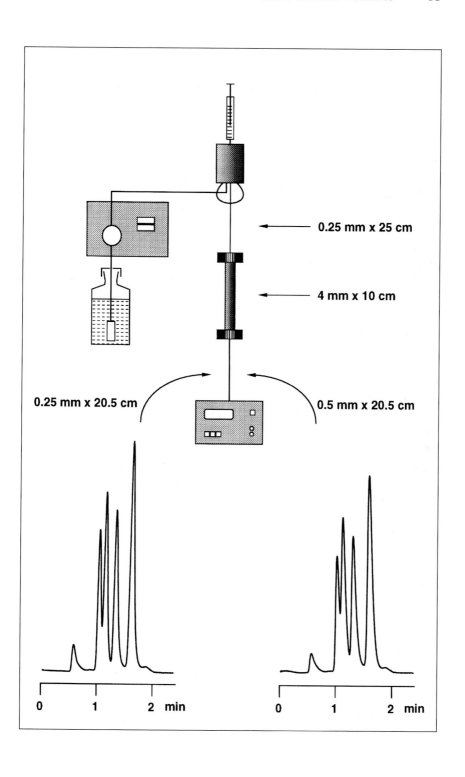

0.25 mm x 25 cm

4 mm x 10 cm

0.25 mm x 20.5 cm

0.5 mm x 20.5 cm

0   1   2   min

0   1   2   min

2.21 Dwell Volume

When working with gradient separations, the dwell volume is a parameter which is often not taken into consideration. This is the volume within the HPLC system between the mixing point of the various solvents and the column entrance. Besides the connecting capillaries the dwell volume includes mainly the mixing chamber (with high pressure gradients) or the mixing valve and the pump (with low pressure gradients); depending on instrument design the volumes of fittings and other items also contribute. The sum of all these volumes delays the time until the gradient profile becomes effective at the column entrance; it also smooths the profile as a result of mixing effects.

If a gradient separation is always performed on the same unmodified HPLC system the dwell volume does not show up (with the exception of computer-aided gradient optimization). In fact a surprise will result if the separation needs to be reproduced on another instrument which probably has a different dwell volume, or if the original instrument is modified.

This separation of PTH amino acids was performed with a high-pressure gradient system of 2.8 mL dwell volume; the volume was the sum of mixing chamber, precolumn (scavenger column), and connecting capillaries (left). Without precolumn and with shorter, narrower capillaries the dwell volume dropped to 0.6 mL (right). Now the retention times are shorter and the separation needs less time but the resolution of the peak cluster at the end is much worse than before because the gradient profile was not designed for the modified HPLC system.

Chromatographic conditions:

Sample:	phenylthiohydantoin derivatives of Asn, Asp, Glu, Ala, His, Pro, Val, Ile and Phe, dissolved in water/methanol 9 : 1
Column:	3.2 mm × 25 cm
Stationary phase:	Spherisorb ODS, 5 μm (reversed phase C_{18})
Mobile phase:	gradient from 10% methanol in water to 70% methanol in 4 min, then isocratic, 1 mL min^{-1}
Detector:	UV 254 nm

Reference: V.R. Meyer

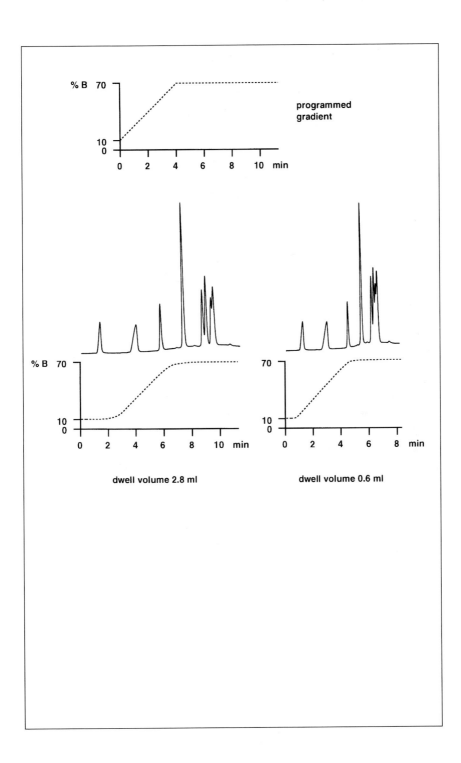

dwell volume 2.8 ml   dwell volume 0.6 ml

2.22 Elution at t_0

Elution at t_0, i.e. at the breakthrough time, is basically unwanted. Compounds which appear that early in the chromatogram have not been retained which means that no chromatography took place. If more than one compound runs through the column this way they cannot be separated. Under such circumstances the certainty that a peak represents a single compound only is much lower than with retained peaks. In addition the standard deviation of peak area (as shown here) or height can be atypically high or low. Quite often the peaks at t_0 have an unusual shape which can make their integration more difficult or easier. The two examples present t_0-peaks of exceptionally poor (top) or good (bottom) peak area reproducibility.

Chromatographic conditions:

Top:	means of 8 analyses
Sample:	50 µL of anion standards in water
Column:	4.0 mm × 25 cm
Stationary phase:	Dionex AS4A-SC
	(anion exchanger for ion chromatography)
Mobile phase:	1.8 mM sodium carbonate + 1.7 mM sodium hydrogen carbonate, 2 mL min^{-1}
Detector:	conductivity (after cation-exchange suppressor)

Reference: R. Brügger, Central Laboratory, Blood Transfusion Service, Swiss Red Cross, Bern, Switzerland

Bottom:	means of 8 analyses
Sample:	extract from tablet containing paracetamol, acetylsalicylic acid, and hexobarbital
Column:	3.0 mm × 15 cm
Stationary phase:	Inertsil ODS-3, 5 µm (reversed phase C$_{18}$)
Mobile phase:	1% potassium dihydrogen phosphate pH 3/acetonitrile/ tetrahydrofuran 50 : 25 : 25, 0.8 mL min^{-1}
Detector:	UV 218 nm

Reference: V. R. Meyer

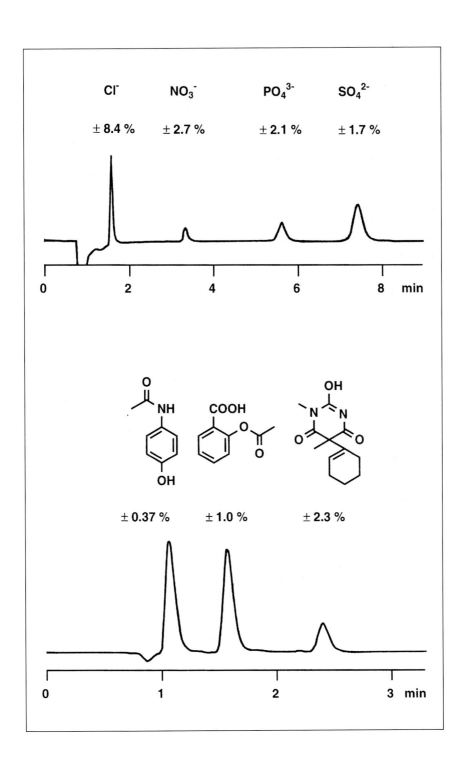

2.23 Classification of C$_{18}$ Reversed Phases

The properties of stationary phases depend on their manufacturing process. The first step is the synthesis of the bare silica; the end-product is influenced by solvents, pH, catalysts, impurities, and the details of the manufacturing process. The material is, therefore, different from one manufacturer to another. The subsequent covalent anchoring of the bonded phase is again subject to widely variable conditions. Thus it is obvious that the commercially available HPLC phases can have very different properties even when, as an example, all belong to the class of C$_{18}$ materials. They differ in pore diameter, pore width distribution, carbon content (the mass fraction of chemically bonded phase), amount of residue silanols and their accessibility as well as in their acidic or basic character.

One can try to classify commercially available C$_{18}$ phases according to their similarity or their differences. A proposal is presented in the diagram. It was established on the basis of the behavior of the phases with regard to hydrophobicity, silanol activity, metal trace activity, and geometric shape selectivity. Adjacent phases are very similar; it would, e.g., be possible to replace YMC ODS by Inertsil ODS. Phases which are far apart show very different properties; it is, e.g., not recommended to use a Spherisorb-type phase for a separation which was developed and optimized on Kromasil.

Depending on the methods of investigation the classification could also look different. For practical purposes it is important to know that the differences exist and that they can be distinct ($\rightarrow$ 2.24).

These considerations are true not only for C$_{18}$ phases but for unmodified silica and for all types of bonded phases as well.

Reference: B. A. Olsen and G. R. Sullivan: J. Chromatogr. A 692 (1995) 147

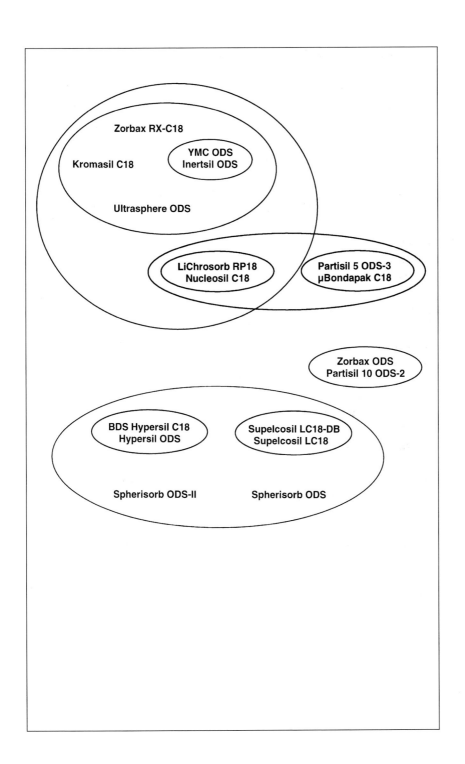

2.24 Different Selectivity of C_{18} Reversed Phases

Because of the different properties of different C_{18} phases ($\rightarrow$ 2.23) it is often not possible to use a stationary phase other than that with which a particular application was developed. Of course, a first attempt can be made with the phase which is at hand and it makes no sense to discourage this approach. A (bad) surprise could, however, result.

This becomes obvious with the presented separations of dirithromycin (an antibiotic) and its degradation products. Retention factors (or retention times), separation factors, resolutions, and elution order can differ markedly. The nice separation on Hypersil ODS looks very different on µ-Bondapak C_{18}. This does not mean that Hypersil is better than Bondapak; perhaps some phases are less suited for certain separations, and the favorable conditions, i.e. the composition of the mobile phase, need to be adapted to the stationary phase. The only consequence of this example is: do not expect to be able to reproduce a published separation on any phase whatever even if it belongs to the same general class.

This is true not only for C_{18} materials but also for all types of HPLC phase. The stationary phase used must be mentioned clearly in the method description ($\rightarrow$ 3.11).

Chromatographic conditions:

Sample:	1 = dirithromycin, 2 = erythromycylamine, 3 = epidirithromycin
Column:	4.6 mm × 25 cm
Stationary phase:	various reversed phases C_{18}, 5 µm
Mobile phase:	50 mM potassium phosphate pH 7.5/acetonitrile/ methanol 37 : 44 : 19, 2 mL min^{-1}
Temperature:	40 °C
Detector:	UV 205 nm

Reference: B. A. Olsen and G. R. Sullivan: J. Chromatogr. A 692 (1995) 147

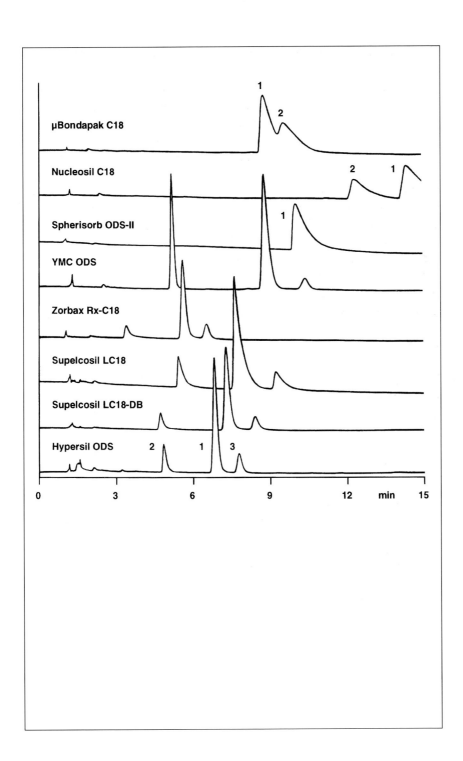

2.25 Different Batches of Stationary Phase

When working with difficult separation problems it is possible that the stationary phase used will not be available with identical properties over a long period of time. In fact even the test chromatograms of the different batches can be almost identical and the difference will only be seen when a critical sample is injected.

The upper separation has more features and is more detailed than the lower one, especially over the second half of the chromatogram. The authors of the study mentioned that only approximately 5% of the columns tested behaved in the desired manner.

In principle only one recommendation can be given for such problems – buy as many columns of a good batch as will be needed to perform the analyses during the next few years. Perhaps another stationary phase would be more rugged in this regard but every change of an optimized difficult separation leads to much work, and it takes a long time to find out about batch reproducibility (in fact, certainty in these matters is an impossibility).

Chromatographic conditions:

Sample:	tryptic peptides from biosynthetic human growth hormone
Column:	4.0 mm × 25 cm
Stationary phase:	LiChrosorb RP-18, 5 μm (reversed phase C_{18})
Mobile phase:	water/acetonitrile with 0.1% trifluoroacetic acid, 1 mL min⁻¹, gradient from 0 to 50% acetonitrile in 60 min
Temperature:	45 °C
Detector:	UV 215 nm

Reference: B.S. Welinder, H.H. Sørensen, K.R. Hejnæs, S. Linde and B. Hansen, in: M.T.W. Hearn, HPLC of Proteins, Peptides and Polynucleotides, VCH, New York 1991, pp. 495–553, Figs. 15–22

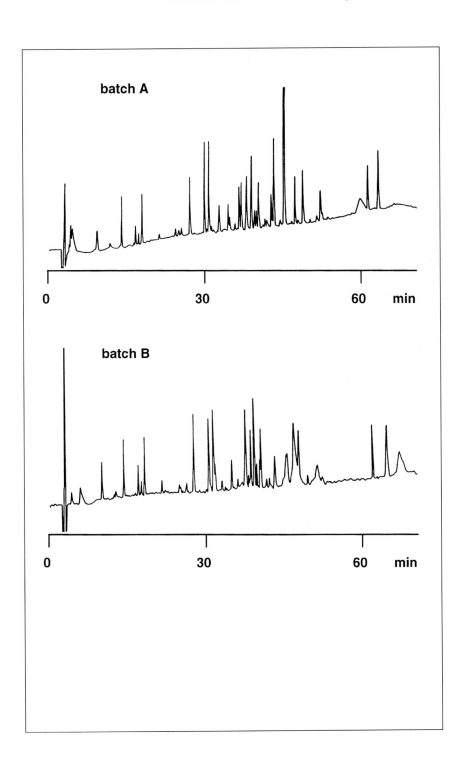

2.26 Chemical Reaction within the Column

Under exceptional circumstances a chemical reaction can occur in the HPLC column. As in the case of decomposition in the sample solution ($\rightarrow$ 2.12, 2.13) the relevant peak decreases in area and an additional peak can appear.

It can be difficult to recognize that a reaction is taking place during chromatography, although it is possible to utilize the general properties of chemical reactions – longer residence time (lower flow rate) or higher temperature should increase the extent of the observed phenomena.

During the separation presented here hydroquinone was partially oxidized to benzoquinone. The oxidizing agent was obviously Fe^{3+} which was present at sufficient concentration in an old, often used silica column. On a new column with identical packing the reaction did not occur and a benzoquinone peak was not found. The reaction was enhanced by dissolved oxygen in the mobile phase.

Chromatographic conditions:

Sample:	hydroquinone (last peak), 1-phenyloctane and diethyl phthalate (first and second peaks)
Column:	4.6 mm × 30 cm
Stationary phase:	Spherisorb S10W, 10 μm (silica)
Mobile phase:	hexane/2-methyl-2-propanol 92 : 8, 0.82 mL min^{-1}
Detector:	UV 270 nm

Reference: C. Y. Jeng and S. H. Langer: J. Chromatogr. Sci. 27 (1989) 549

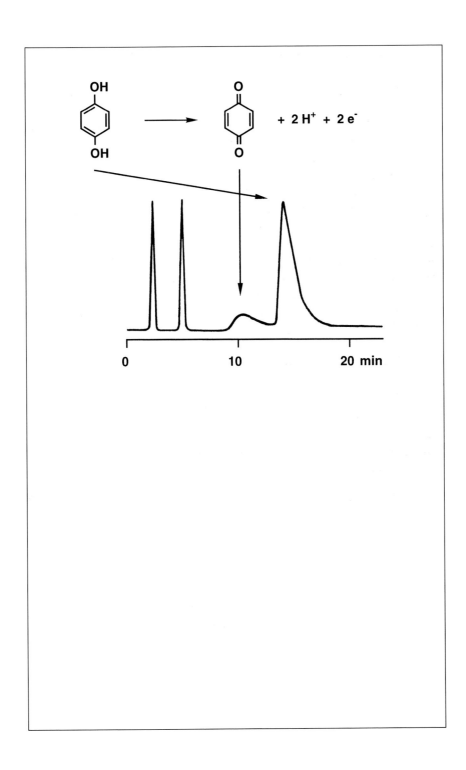

2.27 Recovery and Peak Shape Problems with Proteins

Surprising (and unwanted) behavior of the analytes is not uncommon in HPLC separations of proteins: changes of conformation, loss of activity, and poor recovery. It is not straightforward to explain the observations since there is a great variety of possible interactions of a protein with the stationary phase and of the various possibilities for unfolding and other shape effects.

In general, such problems are less pronounced in hydrophobic interaction chromatography than in ion-exchange chromatography. Nevertheless, very puzzling phenomena can also be observed in HIC as shown here with the separation of cytochrome c. A decent peak shape is only found at elevated temperature but at the cost of decreased peak area, which means loss of protein. The retention time depends on temperature. The effects could not be explained satisfactorily.

Chromatographic conditions:

Sample:	20 µL with 20 µg of cytochrome c dissolved in starting buffer
Column:	4.6 mm × 3.5 cm
Stationary phase:	HIC Butyl NP, 2.5 µm (non-porous reversed phase C_4)
Mobile phase:	A: 5 mM Tris buffer pH 7.4 with 3 M ammonium sulfate
	B: 5 mM Tris buffer pH 7.4 without ammonium sulfate
Gradient:	from 0 to 100% B in 10 min, 1.0 mL min^{-1}
Temperature:	as indicated
Detector:	UV 280 nm

Reference: S. H. Goheen and B. M. Gibbins: J. Chromatogr. A 890 (2000) 73

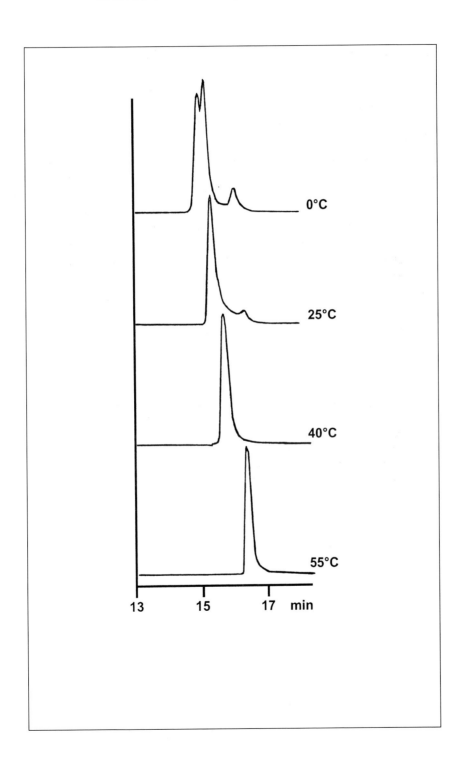

0°C

25°C

40°C

55°C

13   15   17   min

2.28 Double Peaks from Stable Conformers

Some classes of compounds, e.g. proline derivatives, form stable con-
formational isomers in solution with interconversion rates of the same
order of magnitude as the chromatographic analysis time. The stability
of the isomers comes from the amide bond which has some double
bond character, thus giving rise to cis and trans structures.

When captopril, i.e. (S)-1-(3-mercapto-2-methyl-1-oxopropyl)-L-
proline is analyzed in aqueous neutral solution two peaks are obtained
which belong to the cis and trans isomers. In order to obtain a single
peak the mobile phase needs to be acidic in the region of pH 2. Then the
interchange of conformation is rapid compared with the residence time
in the chromatographic phase system. The peak shape can be further
improved by increasing the temperature to 30 °C.

A by-product of captopril is the dimer; with some formulations,
therefore, even three chromatographic peaks are found.

Double peaks from analogous isomers can also be found during
the analysis of carbohydrates on common HPLC phases not especially
designed for sugar separations. In this case the two peaks represent the
two possible anomeric forms of the molecules.

Chromatographic conditions:

Sample: captopril dissolved in mobile phase
Column: 3.0 mm × 15 cm
Stationary phase: Inertsil ODS-3, 5 μm (reversed phase C_{18})
Mobile phase: water/acetonitrile 88 : 12, without additive or at pH 2
                   with phosphoric acid, 0.8 mL min^{-1}
Detector: UV 220 nm

Reference: V. R. Meyer, after U. D. Neue, D. J. Phillips and
           M. Morand: Waters (Europe) Column, Spring 1995, p. 11

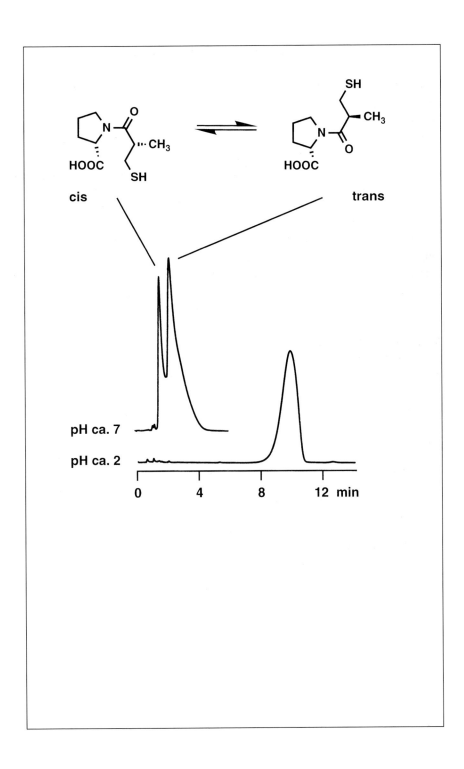

2.29 Influence of Temperature on the Separation

The partition equilibria on which chromatography is based ($\rightarrow$ 1.1) are temperature-dependent. Both the degree and sign of the temperature behavior are specific for any compound, which means that the separation can be influenced by appropriate thermostatting. Without a knowledge of the thermodynamic data it is, however, impossible to predict whether an increase or a decrease in temperature would be advantageous for a given problem. If a separation is optimized for routine analysis it is recommended in any case to investigate the parameter temperature also.

If the analysis is performed without thermostatting one should not be surprised by changes in the chromatogram which arise as a result of fluctuations in ambient temperature during a day or a year.

The example presents an extremely temperature-sensitive separation where a difference as small as 1 °C influences the retention times and even the elution order. In fact this is an exceptionally demanding analysis (isomers of β-carotene) which needs an exceptional phase system: alumina and hexane with a strictly controlled water content.

Chromatographic conditions:

Sample:	isomers of β-carotene
Column:	7.7 mm × 25 cm
Stationary phase:	Spherisorb A5Y, 5 µm (alumina)
Mobile phase:	hexane with defined water content
Temperature:	21 or 22 °C
Detector:	VIS 420 nm

1 = 13,13'-di-*cis*, 2 = 9,13,13'-tri-*cis*, 3 = 9,13'-di-*cis*, 4 = 15-*cis*, 5 = 9,13-di-*cis*, 6 = 13-*cis*, 7 = 9,9'-di-*cis*, 8 = all-*trans*-β-carotene. 2a, 3a, 3b and 3c are not stable and their structure was not elucidated.

Reference: M. Vecchi, G. Englert, R. Maurer and V. Meduna: Helv. Chim. Acta 64 (1981) 2746

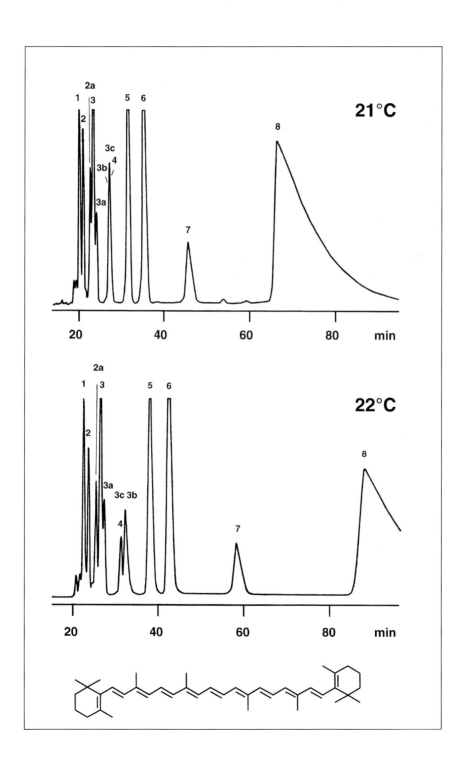

2.30 Thermal Non-Equilibrium within the Column

Eluent viscosity and analyte diffusion coefficients are temperature dependent. Therefore it is important that there are no radial and lateral temperature gradients within the column (or almost no gradients; some temperature rise over the column length is inevitable because of friction heat which is generated by the flow resistance of the column packing). Temperature gradients will occur if the mobile phase has a different temperature than that of the column when it enters the chromatographic bed. This is not the case when both the solvent reservoir and the column are at ambient temperature. However, if the column is heated or cooled it is necessary to thermostat the eluent to the same temperature. If this is not done there will be different flow rates and different mass transfer coefficients over the cross-section of the column, which will result in strange peak shapes and even double peaks.

The upper chromatogram shows a separation which was performed at 77 °C without the use of a preheater, giving grossly deformed peaks. For the lower separation, a stainless steel tube of 0.25 mm × 1 m, coiled to a diameter of 10 cm, was installed between the autosampler and the column and clamped against the heating block in the column oven. This procedure led to peaks of excellent shape. One must not forget that the extra-column volume (→ 2.20) and the dwell volume (→ 2.21) were then larger.

Chromatographic conditions:

Sample:	uracil (not retained), nitroethane, 3,5-dimethylaniline, 3-cyanobenzoic acid, and nitrobenzene
Column:	4.6 mm × 15 cm
Stationary phase:	Zorbax SB-C18, 5 µm (reversed phase C_{18})
Mobile phase:	50 mM potassium phosphate buffer pH 2.6/acetonitrile, 1.5 mL min^{-1} gradient from 5 to 65 % acetonitrile in 13 min
Temperature:	77 °C
Detector:	UV 215 nm

Reference: R. G. Wolcott and J. W. Dolan: LC GC Int. 12 (1999) 14
See also: R. G. Wolcott et al.: J. Chromatogr. A 869 (2000) 211

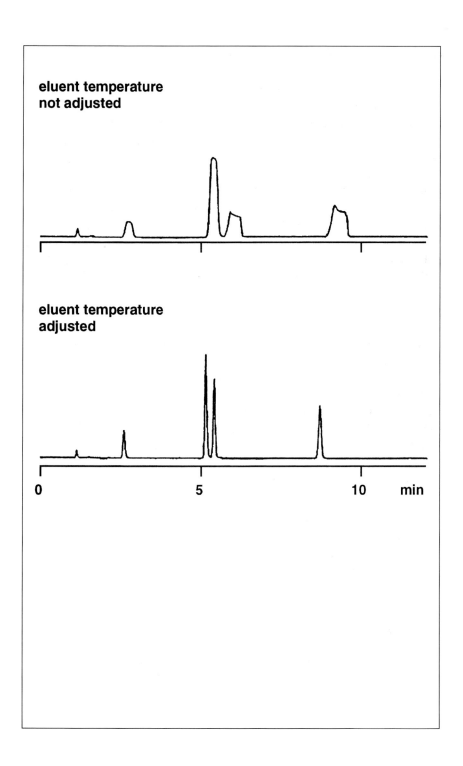

2.31 Influence of the Volume Flow Rate on the Separation

The van Deemter curve ($\rightarrow$ 1.5) illustrates that the height of a theoretical plate depends on the linear flow rate of the mobile phase. This phenomenon becomes more familiar when discussed as the dependence of peak width on the volume flow rate. Its influence on the chromatogram can be remarkably large.

In the figure this is illustrated by a rather unfavorable (or interesting) example, i.e. the incomplete separation of nine PTH amino acids. The peak widths increase with increasing flow rate. This reduces their height. The poorly separated peak pair early in the chromatogram (which probably arises from more than two compounds) loses resolution, which results in increased peak heights. The reason for this behavior, paradoxical at first glance, lies in the overlap of poorly resolved peaks with strong tailing ($\rightarrow$ 2.48).

In most cases chromatographic separations are performed too fast with regard to the van Deemter optimum. The adverse effects on the chromatogram are, however, much less severe than when the flow rate is too low as becomes obvious from study of the van Deemter curve. Especially in reversed-phase chromatography with its rather high-viscosity mobile phases it would be favorable to perform the separations with lower flow rates than usual; a linear flow rate of approximately 0.6 mm s^{-1} is close to the optimum (as calculated for phenol in acetonitrile/water 1 : 1 with $u = vD_m/d_p$ ($\rightarrow$ 1.4), diffusion coefficient $D_m = 1 \cdot 10^{-9}$ m^2 s^{-1}, particle diameter $d_p = 5$ μm). For compounds of higher molar mass the optimum lies at even lower flow rates!

Chromatographic conditions:

Sample:	phenylthiohydantoin derivatives of Asn, Asp, Glu, Ala, His, Pro, Val, Ile and Phe, dissolved in mobile phase
Column:	3.2 mm × 25 cm
Stationary phase:	LiChrosorb SI 60, 5 μm (silica)
Mobile phase:	*tert*-butyl methyl ether
Detector:	UV 254 nm

Reference: V. R. Meyer

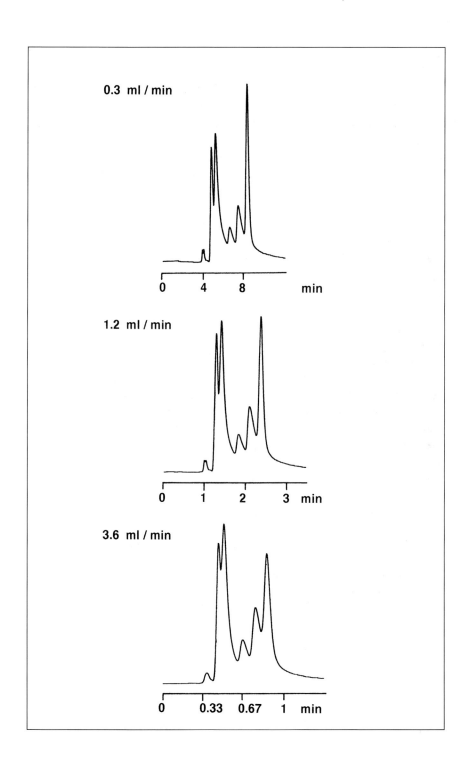

2.32 Influence of Run Time and Volume Flow Rate on Gradient Separations

The separation of a complex mixture can be influenced markedly by both the run time and the mobile phase flow rate. The separation is mainly driven by the increase in the amount of the strong mobile phase; it will (usually) be complete when the final concentration of the B eluent is reached, irrespective of the time necessary for this. The influence of the flow rate is less pronounced. Nevertheless a change of the flow will be apparent from the chromatogram (besides the possible van Deemter effects, → 2.31) because the flowing eluent transports the components, depending on their individual and %B-dependent partition coefficients, to a certain place within the column. The combination of both parameters then gives the chromatogram. If one or the other is changed this will be visible in the peak pattern. Therefore both gradient run time and flow rate (together with the column dimensions), and the dwell volume (→ 2.21) must be noted in the method description (→ 3.11, 3.13).

For the separations presented here it is interesting to note that both combinations, 90 min with 0.3 mL min^{-1} and 60 min with 0.4 mL min^{-1}, give almost identical separations whereas the other chromatograms differ markedly.

In order to conserve solvent one can try to keep the flow rate as low as possible with a given gradient profile (a low dwell volume is advantageous).

Chromatographic conditions:

Sample:	pepsin-digested lactalbumin (48 h, 37 °C)
Column:	2.1 mm × 10 cm
Stationary phase:	Bakerbond Butyl, 7 μm (reversed phase C$_4$)
Mobile phase:	A = 0.1 % trifluoroacetic acid in water
	B = 0.1 % trifluoroacetic acid in water/acetonitrile 2 : 8
Detector:	UV 210 nm

Reference: P. von Haller, Department of Chemistry and Biochemistry, University of Bern, Switzerland

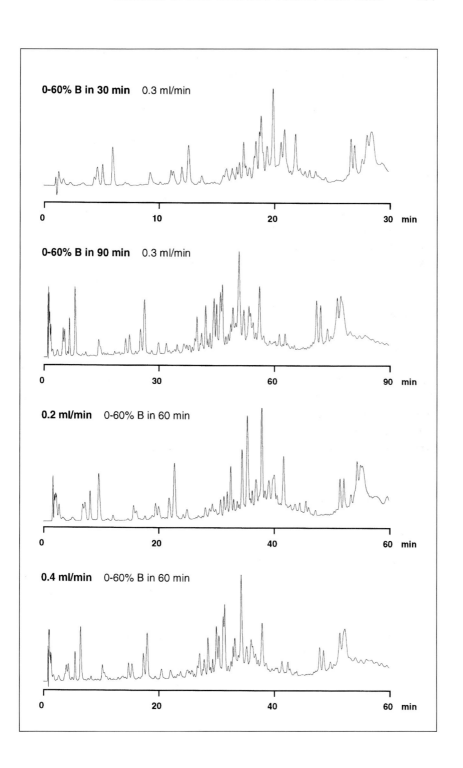

2.33 UV Spectra and Quantitative Analysis

The intensity of a chromatographic peak depends on the injected sample amount, the column properties (dimensions and number of theoretical plates), and the spectroscopic properties of the analyte. With UV detection, an intense signal can be expected if the compound has a high molar absorptivity and if the wavelength of maximum absorptivity is chosen. It is often necessary to detect at a wavelength where all the analytes have some absorptivity but which is not the optimum for one of them. (If this is a problem, the diode array detector should be used.) Without careful checking it is not acceptable to perform a quantitative analysis based on peak heights or areas only. In almost every case it is necessary to determine the calibration curves for all the analytes of interest by the injection of standard solutions ($\rightarrow$ 1.14).

Even for positional isomers the UV-VIS spectra can be highly different in shape and intensity. Of the three isomers of an azo dye, the *meta* compound has a markedly lower absorptivity in the 400–600 nm range than that of the *ortho* and *para* isomers. This leads to highly different peak areas (or heights) when identical amounts are injected.

Chromatographic conditions:

Sample:	isomers of 1-(methoxyphenyl-azo)-2-naphthol (red dyes), 17 ng each in mobile phase; 10 ppm solutions for the spectra
Column:	2 mm × 5 cm
Stationary phase:	YMC-Pack Silica, 3 µm (silica)
Mobile phase:	hexane/dichloromethane 1 : 1, 0.5 mL min^{-1}
Detector:	VIS 500 nm
Peak areas:	*para*, 25.5 %; *meta*, 7.5 %; *ortho*, 67.0 %

Reference: V. R. Meyer

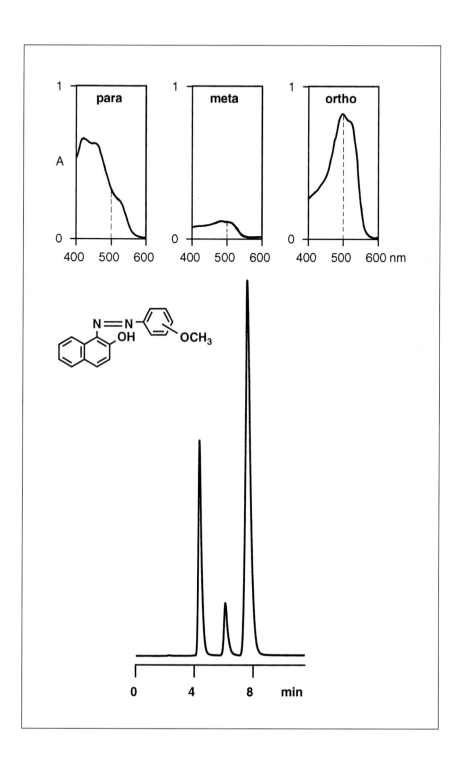

2.34 UV Detection Wavelength

In most cases UV spectra have broad absorption bands which could lead to the assumption that it is of minor importance even for quantitative analyses whether the detector is really working at the desired wavelength. In reality even a deviation of only 1 nm can influence the analytical result although the effect is usually not as severe as in the example shown here.

If naphthalene and anthracene are to be detected quantitatively it is, in fact, very difficult to choose a suitable wavelength. At around 260 nm the absorbance of anthracene depends extremely strongly on the wavelength, which means that a change of not more than 1 nm substantially influences the signal intensity. If identical amounts of the two compounds are injected the anthracene peak is slightly higher than the naphthalene peak at 261 nm under the given chromatographic conditions whereas it is markedly smaller at 262 nm. This analytical method is not rugged ($\rightarrow$ 1.13) with regard to detection wavelength. It would be better to detect at 300 nm, a wavelength where both UV spectra are flat. Unfortunately the detection limit will be poor because of the low absorbance.

Chromatographic conditions:

Sample:	naphthalene and anthracene, 4 µg mL^{-1} each in methanol (also for the UV spectra)
Column:	4.6 mm × 10 cm
Stationary phase:	Kontron RP-8, 10 µm (reversed phase C$_8$)
Mobile phase:	water/acetonitrile 4 : 6, 2 mL min^{-1}
Detector:	UV 261 or 262 nm

Reference: V. R. Meyer

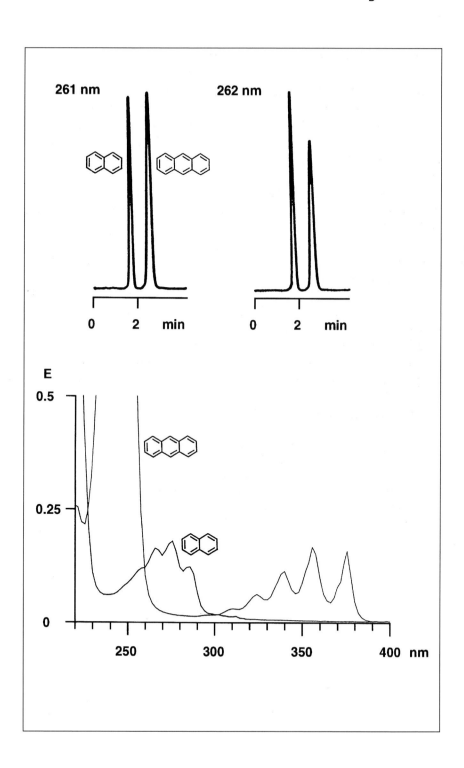

2.35 Fluorescence Quenching by Air

Air (i.e. oxygen) in the mobile phase quenches the fluorescence emission signal: the amount of emitted light is lower than it is in the absence of oxygen. It is therefore necessary to degas the mobile phase thoroughly when a fluorescence detector is used. Otherwise the signal-to-noise ratio ($\rightarrow$ 2.40) is worse or the peak size determination for quantitative analysis can be inaccurate.

Usually the air content of the eluent is unintended. For one of the separations shown here the air was added by sparging because the authors wanted to study the problem in detail. Therefore the quenching effect can be expected to be lower than that found here because the mobile phase is usually not saturated with air. However, without degassing the situation is out of control. For this separation of A vitamins, the intensity of the signals in the air-sparged eluent was approximately two thirds of that using helium-sparged conditions.

Chromatographic conditions:

Sample:	vitamin A esters (13-*cis*-A-palmitate, all-*trans*-A-palmitate, all-*trans*-A-acetate)
Column:	4.6 mm × 15 cm
Stationary phase:	Alltech Econosphere Silica, 3 μm (silica)
Mobile phase:	hexane with 3% diisopropyl ether, 0.04% acetic acid and 0.02 mg mL^{-1} α-tocopherol, 1.5 mL min^{-1}
Detector:	fluorescence 330 nm/470 nm

Reference: A. S. van Brakel, A. R. Matheson and A. K. Hewavitharana:
J. Chromatogr. Sci. 35 (1997) 545

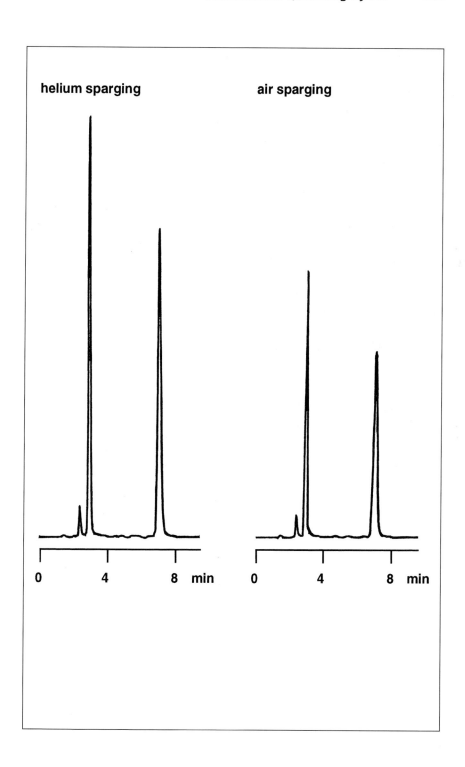

2.36 Detector Overload

The detector response curve ($\rightarrow$ 1.16) shows that the signal of an analyte is too small if detection takes place above the linear range. The analytical result is inaccurate to a lower or higher degree depending on whether the signal is still within the dynamic range or beyond it. To avoid this it is necessary to dilute standard and sample solutions adequately.

The example shows the separation of the enantiomers of 1-(1-naphthylethyl)propylamide on a chiral stationary phase. For the diluted solution (left) the integrator found an area ratio of 79.2 : 20.8. If a concentrated solution of the same mixture of enantiomers is injected (with attenuated detector signal) an unusual shape is observed for the large peak (right); it is not narrow but much too broad. The shape is also not of the type such as arises from severe tailing. In principle the peak should be much higher but the detector cannot register its true height and, therefore, neither its true area because the large mass of compound within the detector cell absorbs the light almost totally. The integrator calculates an area ratio of 71.7 : 28.3, a severe deviation from the accurate value found previously.

Chromatographic conditions:

Sample:	(R,S)-1-(1-naphthylethyl)propylamide dissolved in mobile phase
Column:	4.6 mm $\times$ 25 cm
Stationary phase:	Chiral (R)-DNBPG, 5 µm (dinitrobenzoylphenylglycine, chiral phase)
Mobile phase:	hexane/isopropanol 8 : 2, 1 mL min^{-1}
Detector:	UV 254 nm

Reference: V.R. Meyer

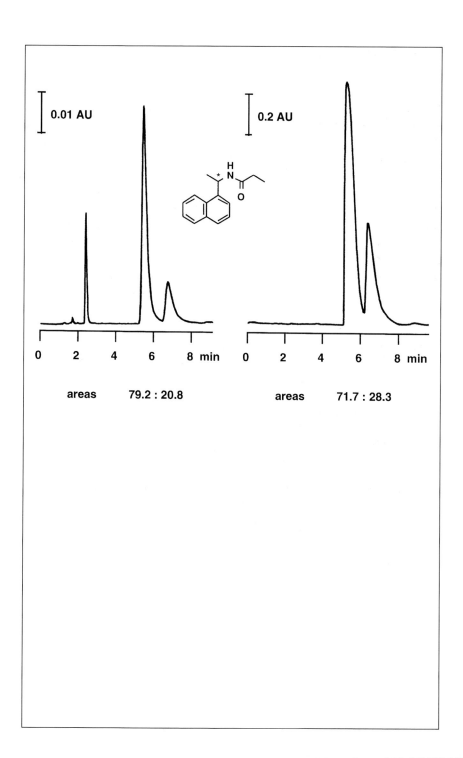

2.37 Influence of the Retention Factor on Peak Height

Under isocratic conditions the later a peak is eluted the broader it is, i.e. the higher its retention factor k. As a consequence its height decreases if the amount of sample remains constant. The k value depends on the composition of the mobile phase and increases with decreasing eluent strength, e.g. in a reversed-phase system with decreasing amount of organic solvent.

This means that the peak height depends strongly on the composition of the mobile phase whereas the peak area is not influenced (although exceptions are possible in very special cases ($\rightarrow$ 2.6)). For quantitative analyses based on peak heights a constant eluent composition must be guaranteed. This is also true for gradient separations: now the gradient profile needs to be reproduced exactly ($\rightarrow$ 2.42).

Some reasons for fluctuations in mobile phase composition:
– The solvent with the lower boiling point evaporates from a bottle of mixed eluent.
– In a gradient system (which can also be used for isocratic separations) the mixing ratio is not (or is no longer) correct because of instrument problems.
– When mixing solvents with pronounced volume contraction, such as methanol and water, the analysts do not always follow the same procedure ($\rightarrow$ 2.1). The best method is to measure both components individually and to mix them afterwards in the reservoir flask.

Chromatographic conditions:

Sample:	10 µg of phenol dissolved in water/methanol 1 : 1
Column:	3.2 mm × 25 cm
Stationary phase:	Spherisorb ODS, 5 µm (reversed phase C_{18})
Mobile phase:	water/methanol, 1 mL min^{-1}
Detector:	UV 254 nm

Reference: V. R. Meyer

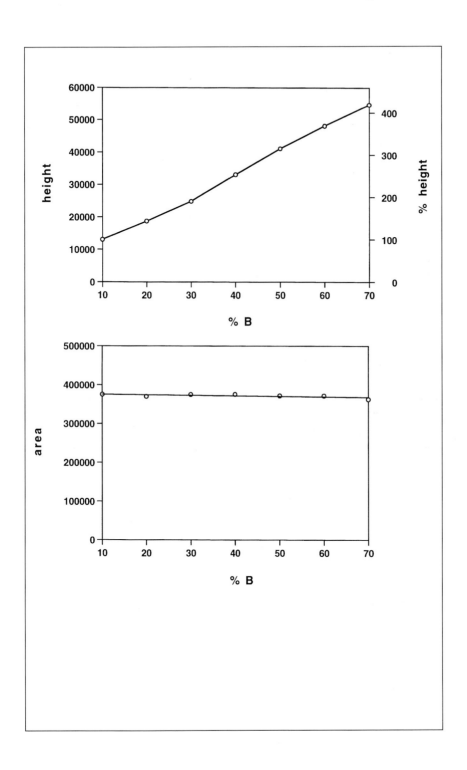

2.38 Influence of the Volume Flow Rate on Peak Area

Most detectors used in HPLC measure the concentration of the analyte in the detector cell – they are concentration-sensitive. The most common example is the UV detector whose signal follows the Lambert-Beer law:

absorbance = concentration × molar absorptivity × optical path length.

(There are also mass-sensitive detectors in use such as the flame ionization detector of GC; these measure a mass flow, in this case the number of ions per unit time.)

When a peak reaches the cell of a concentration-sensitive detector its width is equal to its residence time whereas its height is governed by the analyte concentration at the peak maximum. These two parameters are independent of each other. As a consequence, the peak area depends very strongly on the flow rate of the mobile phase. For quantitative analyses based on peak areas high flow constancy of the pump must be guaranteed.

As can be seen from the lower graph the peak height also depends slightly on the flow rate. This is a result of the dependence of the separation performance on the linear flow rate ($\rightarrow$ 1.5). The higher the theoretical plate number of a column the narrower, and thus higher, the peak will be. Here a flow rate of 0.3 mL min^{-1} was still above the van Deemter minimum (the reduced flow rate, v, was approximately 7).

Chromatographic conditions:

Sample:	10 µg of phenol dissolved in mobile phase
Column:	3.2 mm × 25 cm
Stationary phase:	Spherisorb ODS, 5 µm (reversed phase C$_{18}$)
Mobile phase:	water/methanol 1 : 1
Detector:	UV 254 nm

Reference: V. R. Meyer

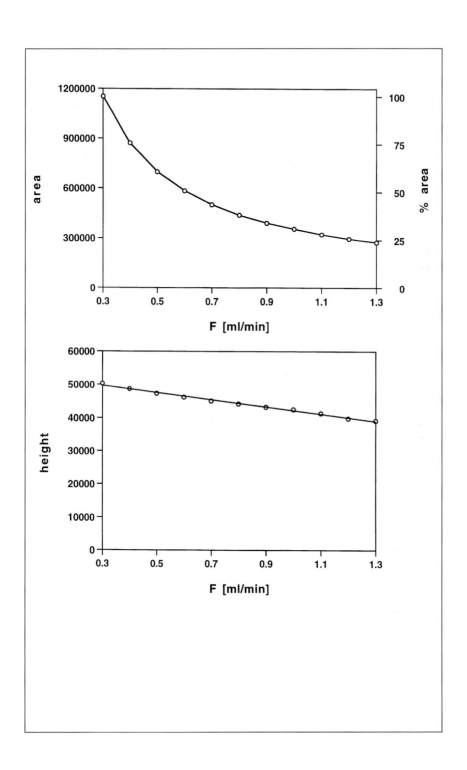

2.39 Leaks in the HPLC Instrument

Leaks in the instrument can influence the chromatogram by multifarious and surprising means. Of course nobody works intentionally with leaky fittings but this does not diminish the danger of this source of error.

Here at three different positions in the instrument a leak of approximately 50% was created, i.e. the flow in each was reduced from 1 mL min^{-1} to 0.5 mL min^{-1} by loosening the appropriate fitting. Such large leaks provoke distinct changes in the chromatogram, which illustrate the effects. With a small leak the consequences are less severe yet large enough to lead to inaccuracies in retention times, peak heights, peak widths or peak areas.

A: No leak.
B: Leak between pump and injector. The volume flow rate is reduced which gives broader peaks ($\rightarrow$ 2.38). The present situation with regard to the van Deemter curve ($\rightarrow$ 1.5) gives rise to increased peak heights. The peak areas increase.
C: Leak between injector and column. The change of the flow rate leads to the same effects as described under B but half of the sample is lost. Therefore the peak heights decrease whereas the peak areas remain unchanged.
D: Leak between column and detector. In principle the chromatogram remains unaltered because the detector measures the concentration and not the mass. Solely the early-eluting, narrow peaks become broader and thus less high because their volume is now too small in relation to the detector cell volume (note the peak height ratios!).

Chromatographic conditions:

Sample:	thiourea, acetone, and methy ethyl ketone dissolved in mobile phase
Column:	4.0 mm $\times$ 10 cm
Stationary phase:	Nucleosil 5 C18, 5 μm (reversed phase C$_{18}$)
Mobile phase:	water/methanol 7 : 3, 1 mL min^{-1}
Detector:	UV 280 nm

Reference: V. R. Meyer, see also V. R. Meyer: J. Chromatogr. A 767 (1997) 25

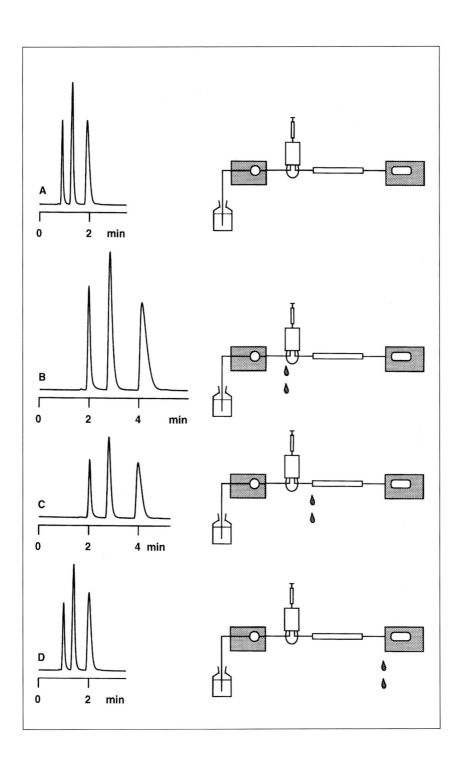

2.40 Impairment of Precision as a Result of Noise

It has already been mentioned that for quantitative chromatographic analysis a signal-to-noise ratio S/N of 10 or more is usually required ($\rightarrow$ 1.17). In a study it was, however, found that S/N must be higher than 50 if a relative standard deviation (= standard deviation/mean) of the peak areas of less than 2% was required (other sources of error which would also affect the precision could be excluded in this study). Although these data were obtained for a particular analytical problem, the determination of chlorthalidone, it can be assumed that the assertion is generally valid. The curve presented in the graph probably matches many HPLC methods, which means that quantitative analysis should not be attempted if S/N is lower than 50 (otherwise the number of injections per sample should be increased). Only the lower x axis of the graph, which gives the amount of sample injected, is not generally valid.

Chromatographic conditions:

Sample:	chlorthalidone ($k \approx 2$), sample volume 10 µL
Autosampler:	Perkin-Elmer ISS-200
Column:	4.6 mm × 33 cm
Stationary phase:	Reduced Activity C18, 3 µm (reversed phase C_{18})
Mobile phase:	water/methanol/acetic acid 65 : 34 : 1, 1.5 mL min^{-1}
Temperature:	30 °C
Detector:	UV 235 nm
Data:	relative standard deviations from 8 analyses each

Reference: M. W. Dong: Views, Perkin-Elmer Newsletter, Fall 1994, p. 7

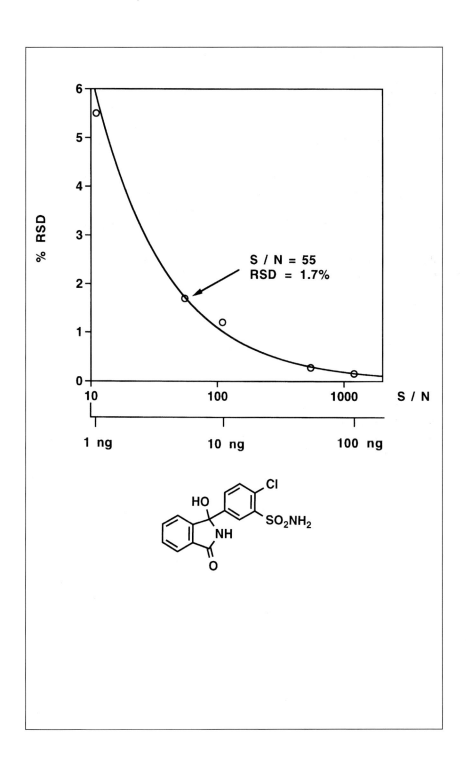

2.41 Determination of Peak Area and Height at High Noise

Integrators and data systems can be misleading as they perform quantitative analysis always with peak areas, whereas the determination of peak heights is often less prone to error (the most important exception being inconstancy of k values, $\rightarrow$ 2.37). For low signal-to-noise ratios the determination of peak heights is more precise.

The example shows a separation with such high noise that the precision is not satisfactory for analysis of the large peak and is unacceptably low for the small peak. The signal-to-noise ratios are approximately 17 for the (R)-amide and 3 for the (S)-enantiomer. In both cases height determination is markedly more precise (or less imprecise) than area determination and is, therefore, recommended. The data obtained for the areas fit well into the curve of M.W. Dong which gives the relationship between the relative standard deviation and the signal-to-noise ratio ($\rightarrow$ 2.40).

Chromatographic conditions:

Sample:	(R,S)-1-(1-naphthylethyl)propylamide dissolved in mobile phase
Column:	4.6 mm × 25 cm
Stationary phase:	Chiral (R)-DNBPG, 5 µm (dinitrobenzoylphenylglycine, chiral phase)
Mobile phase:	hexane/isopropanol 8 : 2, 1 mL min^{-1}
Detector:	UV 254 nm
Integrator:	Hewlett-Packard 3390 A
Data:	means and standard deviations from 5 analyses

Reference: V. R. Meyer

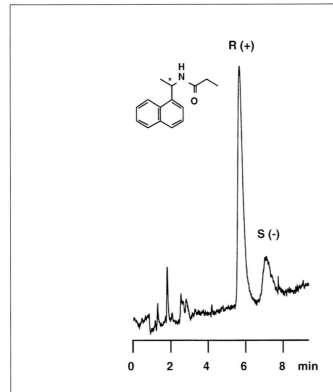

Peak	S / N	Area *	RSD **	Height *	RSD **
R (+)	17	69320 ± 5020	7.2 %	2820 ± 120	4.3 %
S (-)	3	12950 ± 7390	57 %	480 ± 110	23 %

*   integrator counts

**   relative standard deviation

2.42 Peak Height Ratios

It seems to be a matter of course for quantitative analyses that a calibration curve is established. There is in principle just one single exception to this rule: Enantiomer ratios may be determined without calibration if the chiral compounds are not derivatized to diastereomers and if quantitation is performed by peak area. The procedure can be used for isocratic and gradient separations as well. (If, however, high accuracy is needed it is necessary to check this approach carefully with calibration standards.)

In contrast to peak area determinations it is not permissible to perform quantitative analyses without a calibration curve if peak heights are used, not even with enantiomers. Under isocratic conditions the peaks become broader with increasing retention time, and thus also less high, as shown in the chromatogram to the left. The effect can be less pronounced than in the example but it is always present. With gradient separations it is not possible to predict the peak height ratio; the second peak of a racemate can be smaller, equally high by chance or even higher, depending on relative retention and gradient profile. For the chromatogram to the right the gradient step was positioned intentionally at the appropriate retention time, which caused the second peak to be narrower and, therefore, also higher than the first.

Thus peak height determination for quantitative analysis is only possible in combination with an accurate calibration curve, even for enantiomers.

Chromatographic conditions:

Sample: racemic 1-(1-naphthylethyl)propylamide and toluene
                   (t_0 peak)
Column: 2.1 mm × 12.5 cm
Stationary phase: Nucleosil 5 NH$_2$ with ionically bound (R)-DNBPG,
                   5 μm (dinitrobenzoylphenylglycine, chiral phase)
Mobile phase: hexane/isopropanol, 0.5 mL min^{-1}
                   left: 95 : 5 isocratic
                   right: step gradient from 5 to 10% isopropanol after
                   10 min
Detector: UV 254 nm

Reference: V. R. Meyer: J. Chromatogr. 623 (1992) 371

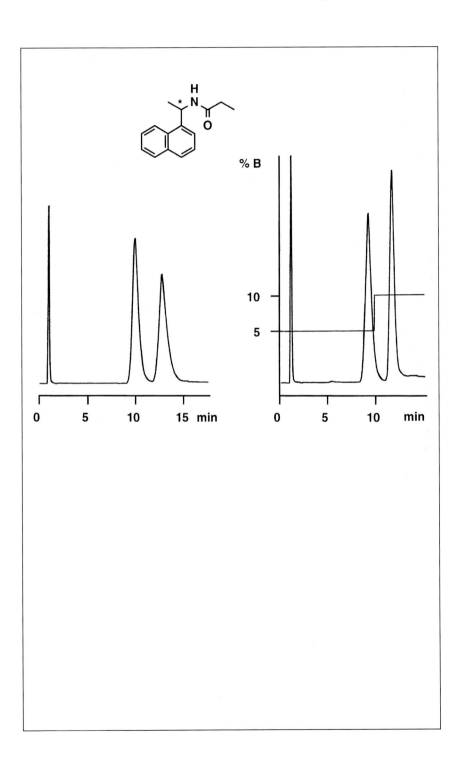

2.43 Incompletely Resolved Peaks

Incompletely resolved peaks always give rise to inaccurate area integration. This error is often ignored or underestimated because 'only' the accuracy is affected but not the precision, and therefore the analyst does not become aware of it.

The graph shows two pairs of overlapping peaks. The contours of the individual peaks are visible, which is not the case with real chromatograms. In reality both man and computer only see the sum curve which is drawn here with a bold line. During the integration the data system identifies the lowest point between the peaks and divides them (usually) with a vertical drop. The areas left and right of this line belong then to the left and right peaks, respectively. From the drawings it becomes clear that this area separation is wrong, with only a few special cases being exceptions. Let us look at the situation with regard to the second peak:

With a vertical drop the second peak wins the additional area which is marked in black, but it loses the hatched part. If the peaks have Gaussian (symmetrical) shape the hatched area is larger than the black one and the total area is too small. In the case of tailing, which is common in chromatography, the relationship is reversed and the integrated area is too large.

The large peak also is not integrated accurately; in the upper chromatogram its area is too large, in the lower one it is too small. Yet its relative error is lower; the absolute areas which belong to the wrong peak are identical for both peaks (with opposite sign) but the effect is less severe for the large one.

These errors are purely geometrical in nature and do not depend on the integrator or data system used. Peak deconvolution software packages are commercially available but in any case it is better to optimize a separation for high resolution than to perform correcting calculations afterwards.

Conditions:

Gaussian:	area ratio 3 : 1, resolution 1.0. Area of the small peak is 97% of true
Tailing:	a tailing b/a of 1.5 ($\rightarrow$ 1.2) was overlaid on the same peaks. Area of the small peak is 115% of true

See also: V. R. Meyer: LC GC Int. 7 (1994) 590

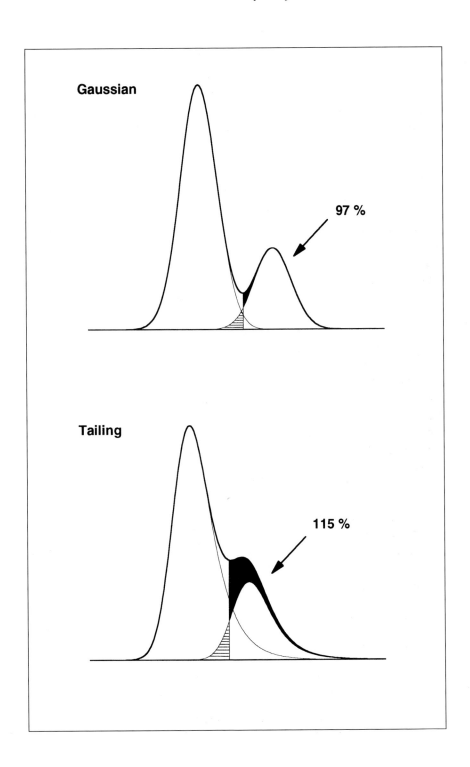

2.44 Area Rules for Incompletely Resolved Peaks

The errors which occur in the determination of the areas of incompletely resolved peaks can be described qualitatively in a very simple manner. No more than the following rules need to be known:

With Gaussian (symmetrical) peaks the large peak is too large, the small one is too small, irrespective of the order of elution.

With tailing peaks the first peak is too small, the second one is too large, irrespective of the size ratio.

The ratio of the relative errors is inversely proportional to the peak area ratio, i.e. small peaks are affected more strongly. For an area ratio of, e.g., 10 : 1 the percentage error is ten times greater for the small peak than for the large one.

The errors increase with decreasing resolution and with increasing tailing. The larger the difference between the sizes of the peaks, the greater the relative error becomes for the small peak and the less it becomes for the large one.

Conditions:

Area ratio 1 : 3 and 3 : 1, respectively
Resolution of the Gaussian peaks 1.0
Tailing b/a of the asymmetric peaks 1.5

See also: V. R. Meyer: J. Chromatogr. Sci. 33 (1995) 26

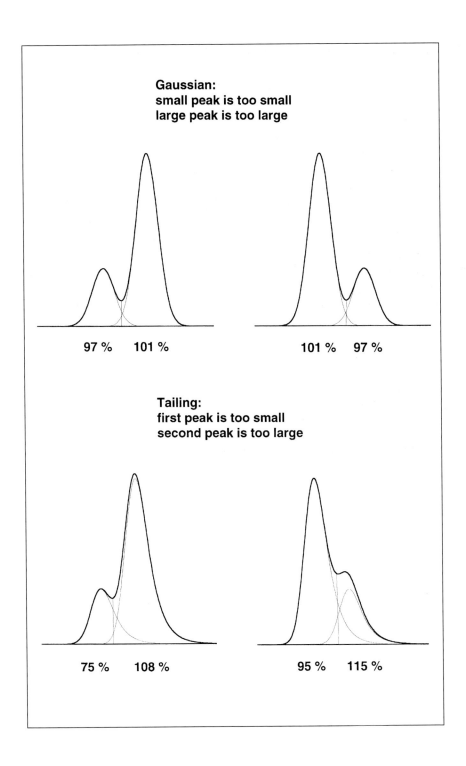

2.45 Areas for a 1 : 10 Peak Pair

The graph presents the areas, as found by an integrator, of the small peaks of a poorly resolved pair of size ratio 1 : 10 or 10 : 1, respectively. On the x axis the resolution of the two peaks runs from 2.0 to 0.8 with the small peak eluted before the large one in the left half of the graph and vice versa in the right half. The small drawings show the pairs with certain resolution, without tailing, however. The y axis indicates the areas found in integrator counts if the peaks are separated by a vertical drop at the deepest point between them. (With another type of integrator the numbers on the y axis would be different but the curves would be identical.) The horizontal line in the middle of the graph shows the true area of the small peak. Different degrees of asymmetry (of both peaks), determined as tailing b/a ($\rightarrow$ 1.2) were investigated. The empty space in the middle of the graph represents the conditions where the integrator was no longer able to differentiate between the two peaks.

As discussed previously the area of the small peak is too small for symmetrical peaks ($\rightarrow$ 2.44); the unbroken line with $b/a = 1.0$ drops from the true area to smaller values with decreasing resolution. With tailing ($b/a > 1.0$) the integrated area is too small if the small peak is eluted first and too large if it is eluted last. It can be seen that even a resolution of 2.0 is not high enough if the small peak lies behind the large one and if the peaks tail, as is common in most HPLC separations.

For the large peak the errors are ten times smaller and have the opposite sign.

For quantitative analysis of peak areas the resolution must be reasonably (!) high. The resolution must be higher the more extreme is the size ratio of the peaks.

Reference: V. R. Meyer: LC GC Int. 7 (1994) 94 or LC GC Mag. 13 (1995) 252

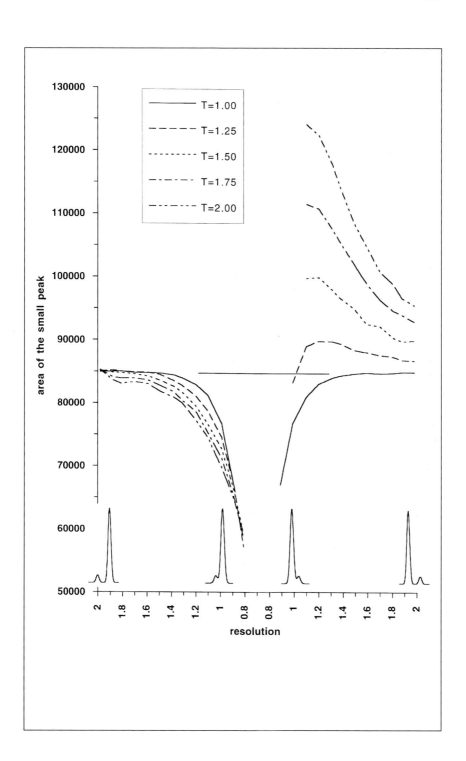

2.46 Heights for a 1 : 10 Peak Pair

For the same peak pair with a size ratio of 1 : 10 or 10 : 1, respectively, the height of the small peak instead of the area ($\rightarrow$ 2.45) is now considered. The integrator separated the two peaks by a vertical drop at the deepest point between them. For this problem no 'generally true' peak height can be indicated because the peak becomes smaller and smaller with increasing tailing if the area remains constant; therefore this graph has no horizontal line in the middle.

It becomes obvious that in the case of tailing (of both peaks although in principle only the shape of the first one is of importance) the small peak must not be eluted after the large one if the resolution is not large enough. 'Enough' means that the baseline must be really reached between the peaks. With stronger tailing better resolution is necessary. The curves on the right side of the graph make clear that 'rider peaks' on the slowly falling trailing edge of a large peak cannot be quantitated accurately; this is also true for their area determination ($\rightarrow$ 2.47). In contrast to this the quantitative analysis is accurate or has only a minor error if the small peak is eluted in front of the large one and if its height is measured. Even with resolution which is at the limit of peak recognition, here as low as 0.8, the error is small.

Reference: V. R. Meyer: LC GC Int. 7 (1994) 94 or LC GC Mag. 13 (1995) 252

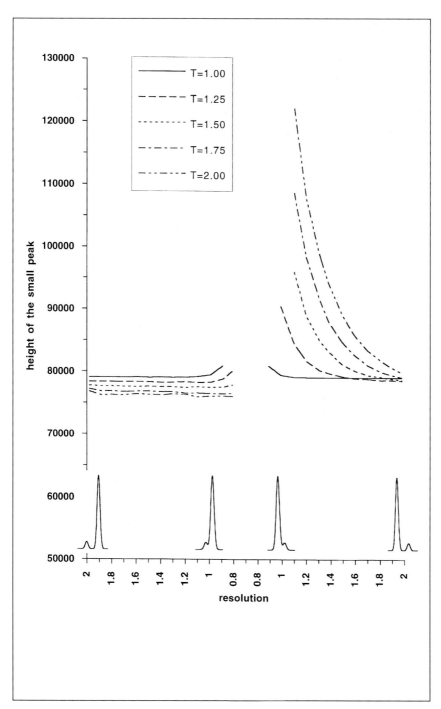

2.47 Quantitative Analysis of a Small Peak

Without calibration under 'real' conditions it is impossible to quantitate a small peak which sits on a large one. 'Real' conditions means here that the standard solutions used for the calibration curve contain all the components in the typical concentrations at which they are present in the sample and which influence the quantitative analysis by peak overlap, as shown here, or by any other effects.

The figure shows the quantitative analysis of 2,3-dimethylphenol, which is eluted on the tail of the 2,3-xylidine peak. Injected as a pure compound (top) 0.8 µg of dimethylphenol gives a peak area of 0.405 mV min and a height of 2.51 mV. In the presence of 80 µg of xylidine (middle) the question arises which method is best for the quantitative analysis of the small rider peak. The peak borders can be drawn by vertical lines or by means of a tangent, the peak size can be determined by area integration or by measurement of height.

It is found that no method gives an accurate result.
Vertical drop, area: 0.893 mV min = 220%
Vertical drop, height: 4.03   mV   = 161%
Tangent, area: 0.333 mV min = 82%
Tangent, height: 2.26   mV   = 90%

With slightly better resolution, which would bring the small peak to a lower position on the xylidine tail, other results would be obtained. Nevertheless it can be stated as a general rule that the vertical drop method, applied to rider peaks, gives values which are too high. In this case even two vertical drops were drawn; an integration down to the peak end would result in a much greater area. The widely used tangent method generally yields values which are too small, not only with this separation problem!

Chromatographic conditions:

Sample:	80 µg of 2,3-xylidine and 0.8 µg of 2,3-dimethylphenol
Column:	4.0 mm × 25 cm
Stationary phase:	LiChrospher 60 RP Select B, 5 µm (reversed phase C_8)
Mobile phase:	water/methanol 33 : 67, 1.5 mL min^{-1}
Detector:	UV 280 nm

Reference: V.R. Meyer, see also V.R. Meyer: Chromatographia 40 (1995) 15; S. Jurt et al.: J. Chromatogr. A 929 (2001) 165

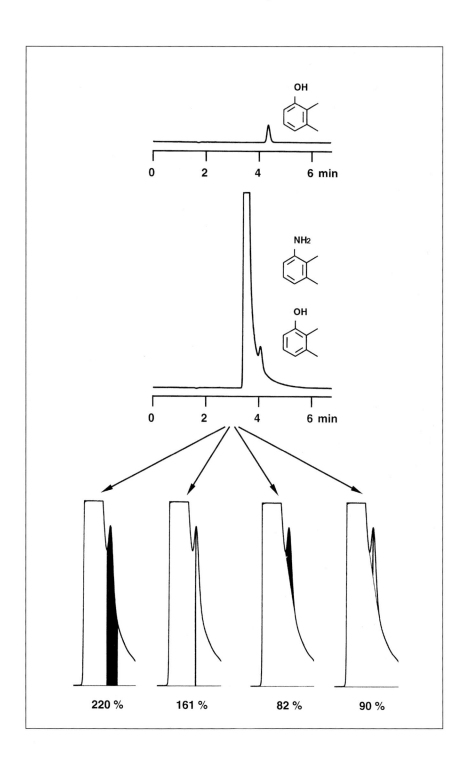

2.48 Incompletely Resolved Peaks with Tailing

Most peaks of real chromatograms are not symmetrical but show some tailing. With inadequate resolution this leads to wrong apparent peak heights. For us observers as well as for the data system only the sum curve is visible and not the individual peak contours.

The graph shows two identical peaks with tailing $b/a = 1.5$ ($\rightarrow$ 1.2) in the upper half. Although the resolution does not go down to the baseline it is good enough to present both signals with identical heights. In the chromatograms below the resolution is much poorer, either because of broader peaks (the theoretical plate number N is smaller) or because of a decrease in the relative retention (separation factor α). The same effect could be observed with a strong increase of the peak tailing, e.g. as a result of column ageing: the second peak is definitely higher than the first one because it sits on its broad tail. The relative heights which were initially 1 : 1 now are 1 : 1.09 in both cases. This ratio depends on the individual circumstances, of course.

Likewise peak areas cannot be determined accurately if the integrator divides the peaks by a vertical drop at the lowest point between them ($\rightarrow$ 2.43, 2.44). The integrated area ratio changes from the initial value of 1 : 1.03 (due to incomplete resolution) to 1 : 1.3 in both cases.

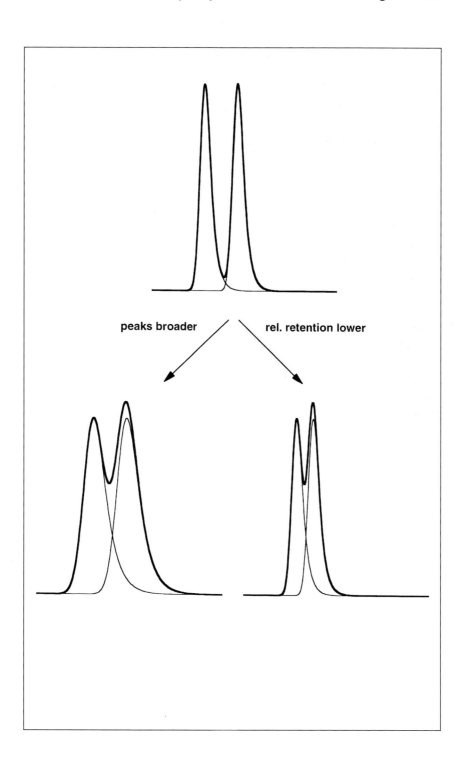

peaks broader   rel. retention lower

2.49 Integration Threshold and Number of Detected Peaks

Not only the detector range ($\rightarrow$ 2.51) but also the integrator threshold can strongly influence the analytical result. By the selection of a certain threshold value the analyst defines the positive deviation from the baseline, which will be recognized by the data acquisition as the starting point of a peak. If the threshold, measured in millivolts, is set high above the baseline, small signals will not be recognized as peaks and will not be integrated. For the analyst it can be difficult to choose the proper threshold: if it is too low, noise "peaks" will also be detected, if it is too high, the small true peaks will be lost. It can be necessary to indicate both the type of data system and the chosen threshold in the standard operating procedure ($\rightarrow$ 3.13).

When analyzing a reactive textile dyestuff (a technical product with many by-products), it was found that the number of peaks which were detected by the data system was strongly dependent on the threshold value. This parameter could be selected on a scale from -12 (sensitive) to $+4$ (not sensitive). With regard to the integration parameters this method is not rugged ($\rightarrow$ 1.13).

Chromatographic conditions:

Sample:	Cibacron Red C-2G in water
Column:	4.0 mm × 12.5 cm
Stationary phase:	ODS-Hypersil, 5 µm (reversed phase C_{18})
Mobile phase:	water/acetonitrile with 28% methanol, 0.4 g L^{-1} tetrabutylammonium perchlorate and sodium citrate, pH 6.2, gradient from 10 to 90% acetonitrile, 1 mL min^{-1}
Detector:	UV 254 nm

Reference: Y. L. Grize, H. Schmidli, and J. Born: J. Chromatogr. A
686 (1994) 1

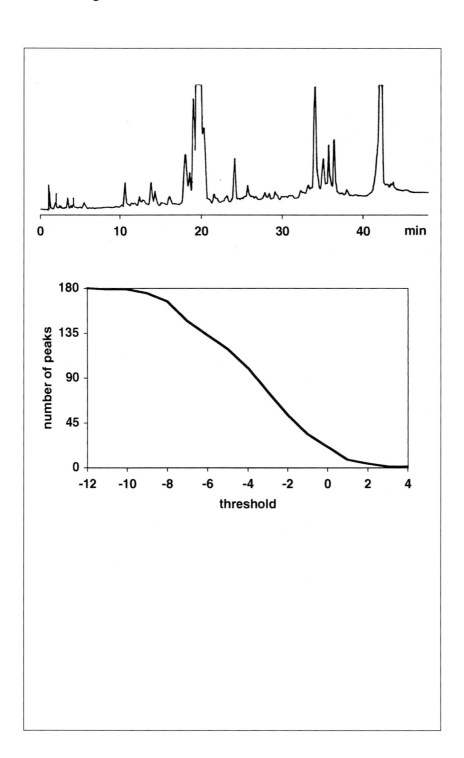

2.50 Detector Time Constant and Peak Shape

The time constant of an electronic device is the time which is needed by an incoming rectangular signal to reach 63% of its height. Electronic data processing distorts all signals because of this inevitable delay. Since a detector is built up from numerous electronic parts some minimum distortion is always happening. In addition, the time constant of a detector used in the laboratory can be selected over a wide range. A small (perhaps too small) time constant allows the true peak shape to be represented as well as the noise. An increase in the time constant distorts the peaks and smooths the noise. Although the peak area remains constant the tailing increases and the peak height decreases.

Peak data for hexylbenzene/octylbenzene:

Time constant s	Retention time	Peak area	Plate number	Asymmetry
0.1	4.14 / 6.80	38026 / 29654	8500 / 8590	0.94 / 0.86
0.5	4.15 / 6.81	38319 / 29977	7930 / 8390	0.98 / 0.86
2	4.17 / 6.84	38604 / 30194	5030 / 6480	1.5 / 1.1
4	4.18 / 6.86	38164 / 29913	3080 / 4610	2.4 / 1.7
8	4.20 / 6.86	37959 / 30089	1450 / 2630	4.2 / 2.8

Chromatographic conditions:

Sample: hexylbenzene and octylbenzene in mobile phase
Column: 4.0 mm × 12.5 cm
Stationary phase: LiChrospher RP-18, 5 µm (reversed phase C_{18})
Mobile phase: water/methanol 90 : 10, 1 mL min^{-1}
Detector: UV 250 nm (Merck LaChrom)

Reference: D. Stauffer, Hoffmann-La Roche AG, Basel, Switzerland

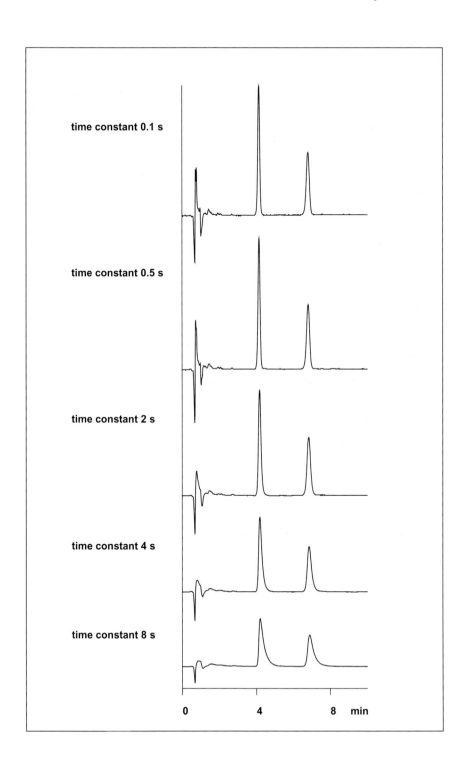

time constant 0.1 s

time constant 0.5 s

time constant 2 s

time constant 4 s

time constant 8 s

0   4   8   min

2.51 Quantitative Analysis in the 99 % Range

For content analyses of, e.g., intermediates from synthesis or pharmaceutical end-products an HPLC separation is often demanded. A common assay is to add all peak areas or heights, as integrated at a certain
wavelength, and to equate the sum to 100 % (being aware that the various compounds have their individual absorbances); the purity of the
analyte as determined by this method must then be higher than 99 % or
the like.

   The height of the small peaks, and thus their detectability, now
depends strongly on external parameters: the detector range chosen,
the integration parameters, and the signal-to-noise ratio, which itself
depends on the amont of sample injected. In addition the main
peak might be so large that it exceeds the linear range of the detector
($\rightarrow$ 1.16, 2.36) and cannot be integrated accurately. The approach is
unfavorable and not recommendable; high-low chromatography is to
be preferred ($\rightarrow$ 3.7). If a data system is used it is, in fact, possible
to set the integration limits manually (with the danger of subjectivity)
after the separation; this is not possible with an integrator.

   For this example two different detector ranges were used whereas
all the other parameters remained unchanged. This gave a main peak
'purity' of either 100 % or 98.9 %. But even the latter content is certainly
not accurate because the peaks overlap ($\rightarrow$ 2.43) and some of the by-
products of lowest concentration are not integrated at all.

Chromatographic conditions:

Sample:	10 µL solution of vitamin B_{12} derivative in mobile phase
Column:	3.2 mm × 25 cm
Stationary phase:	LiChrosorb SI 60, 5 µm (silica)
Mobile phase:	hexane/dichloromethane/isopropanol 8 : 2 : 1, 1 ml. min^{-1}
Detector:	VIS 360 nm, 0.1 and 0.5 AU, respectively
Integrator:	Hewlett-Packard 3390 A

Reference: V. R. Meyer

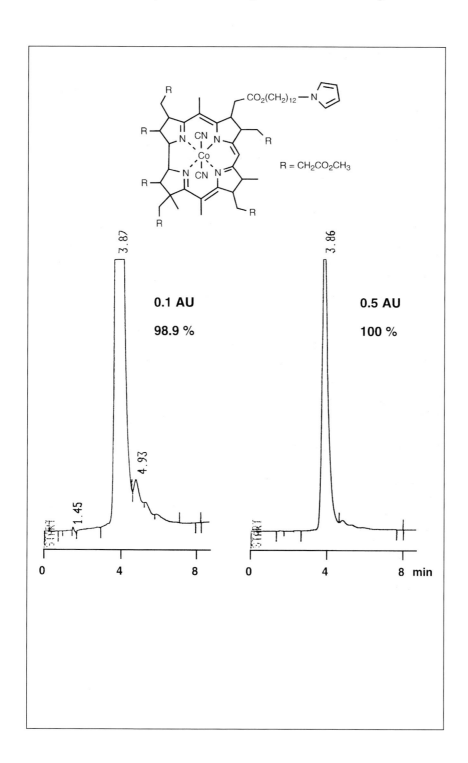

2.52 Correlation Coefficient of Calibration Curves

The correlation coefficient, r, represents how well a calculated calibration curve fits the experimental data points. The maximum value is 1.0; attempts are always made to reach this value. Because the calibration curve should be linear, its usual representation is $y = ax + b$, where x stands for the pre-set data, such as concentrations, and y means the found data, such as peak areas or heights.

It is not advisable to obtain the correlation coefficient from the computer without having a look at the curve. The upper graph shows data with an unambiguous trend which is best described by a quadratic equation (upper right); the function is not straight but curved. However, the deviations are small enough to give a correlation coefficient of 0.999 with linear fit. The lower example shows a data set with commonly observed scatter – some points are too high whereas others are too low, but there is no trend. The correlation coefficient is again 0.999. Now the linear fit is correct ($\rightarrow$ 3.5).

Of course it is not advisable to fit each set of data with a polynomial (of any degree). Before doing so it is necessary to explain the reason for the deviation from linearity. The goal should always be to obtain linear relationships.

Data:

x	y top	y bottom
0		
1	1.2	1.2
2	2.1	1.9
3	3.0	3.0
4	3.9	4.1
5	4.8	4.8
6	5.9	6.1
7	7.0	7.0
8	8.1	7.9
9	9.2	9.2
10	10.3	9.7

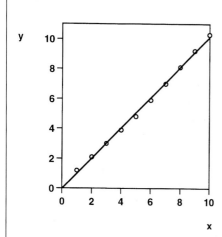

$y = 1.015x - 0.033 \quad r = 0.999$

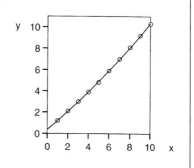

$y = 0.019x^2 + 0.807x + 0.383 \quad r = 1.000$

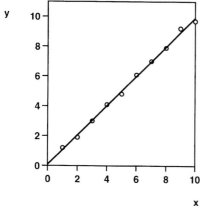

$y = 0.982x + 0.087 \quad r = 0.999$

Part III

Useful Strategies

3.1 Column Tests

After purchase each HPLC column should be tested. According to internal guidelines ($\rightarrow$ 3.12) the test is repeated at regular time intervals and whenever column deterioration is suspected. The test results are kept (either physically or as computer files) for as long as the column is in use.

The test can be performed with standard compounds (under isocratic conditions) or with a typical sample for which the column will be used. With the first type of test it is advisable to calculate some figures of merit whereas the second type is primarily used to judge the quality of the separation with regard to the resolution of a critical peak pair or to the peak shape.

The upper chromatogram shows a test with standard compounds. The calculated figures of merit are: theoretical plate number N ($\rightarrow$ 1.2) of the last peak = 14 500, reduced plate height h ($\rightarrow$ 1.4) of the last peak = 3.0, tailing T ($\rightarrow$ 1.2) of the last peak = 1.3, and reduced flow resistance ϕ ($\rightarrow$ 1.4) = 970.

The lower chromatogram shows the separation of PTH amino acids on the same column. It is known from experience that the separation of 19 amino acids as found here is only possible on an excellent column.

Chromatographic conditions:

Column:	2.1 mm × 22 cm	
Stationary phase:	PTH C-18, 5 µm (reversed phase C_{18})	
Top:	Sample:	benzyl alcohol, benzaldehyde, and benzoic acid methyl ester
	Mobile phase:	water/acetonitrile 6 : 4, 0.4 mL min^{-1}
	Pressure:	184 bar
Bottom:	Sample:	phenylthiohydantoin amino acids
	Mobile phase:	gradient with water, tetrahydrofuran, acetonitrile, isopropanol and additives, 0.21 mL min^{-1}
	Detector:	UV 269 nm

Reference: Applied Biosystems and U. Kämpfer, Department of Chemistry and Biochemistry, University of Bern, Switzerland, respectively

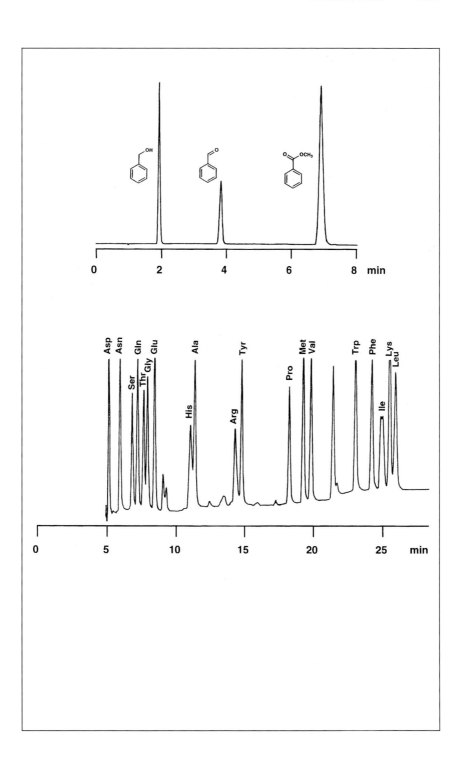

3.2 Apparatus Tests

Accurate, precise, and reproducible analytical results can only be obtained if the various functions of the HPLC system are tested and adjusted regularly. There are no generally mandatory procedures for such tests because every laboratory or company is free to prescribe how often and to what extent they shall be performed. The results need to be noted in a log-book. Binding rules should, moreover, state how to proceed if the specifications are not reached. It is possible to hire a technician of a specialized company for the tests.

A comprehensive test includes the following functions:
- For the pump, the flow accuracy (e.g., better than 5%), the short-term and the long-term flow constancy (e.g., better than 0.5% and 0.2%, respectively).
- For the injector, the reproducibility (e.g., better than 0.5%); it is necessary to distinguish between partial and total filling of the loop ($\rightarrow$ 2.16, 2.17). For the autosampler, the accuracy and the injection carry-over also.
- For the detector, the noise (e.g., less than 0.04 mAU, 1 AU = 1 absorbance unit) and the wavelength accuracy ($\rightarrow$ 2.34, 3.3) (e.g., better than 2 nm) of UV or fluorescence detectors.
- For the gradient system, the accuracy and reproducibility of the profile.

Reference (partially): G. Maldener: Chromatographia 28 (1989) 85
V. R. Meyer: Practical High-Performance Liquid
Chromatography, Chapter 24

Apparatus Tests

Pump:	flow accuracy, short-term flow constancy, long-term flow constancy
Injector:	reproducibility
Autosampler:	reproducibility, accuracy, carry-over
Detector:	noise, wavelength accuracy
Gradient system:	accuracy, reproducibility of the profile

3.3 Wavelength Accuracy of the UV Detector

In a laboratory which works in accordance with a quality assurance system ($\rightarrow$ 3.18) it is necessary to check the wavelength adjustment of the UV detector regularly ($\rightarrow$ 2.34). Such a test is also recommended for all other laboratories. Various test compounds can be used but one with distinct and narrow absorbance maxima is always to be preferred over another with a single broad maximum, as is often found in organic molecules. Benzene (as hexane solution or even better as vapor) could be used for calibration in the UV but because of health considerations this cannot be recommended.

An interesting compound is erbium perchlorate which has several sharp maxima in aqueous solution. They include the important wavelength of 254.6 nm and several bands in the visible region.

Terbium perchlorate with an absorption maximum at 218.5 nm can be used for calibration at low wavelengths.

Reference: B. Esquivel, poster at the 17th International Symposium on
Chromatography, Vienna 1988
J. B. Esquivel: Chromatographia 26 (1988) 321

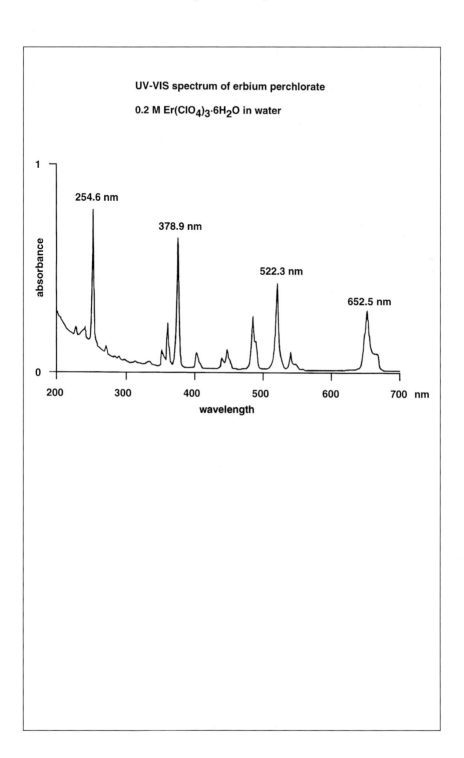

UV-VIS spectrum of erbium perchlorate

0.2 M $Er(ClO_4)_3 \cdot 6H_2O$ in water

254.6 nm

378.9 nm

522.3 nm

652.5 nm

absorbance

wavelength

3.4 Internal Standards

With the correct use of an internal standard all possible problems with sample injection are eliminated. The internal standard must be added in identical concentration to all calibration and sample solutions in order to provide a reference for quantitative analysis; now not peak areas or peak heights are compared but peak-area or peak-height *ratios*.

In HPLC injection, problems are less pronounced than in gas chromatography. Usually internal standards are used to cope with another source of possible problems, i.e. sample preparation. If the proper standard is added to the sample at an early stage, i.e. as the first step of preparation, many error sources with extraction, work-up, filtration, incomplete quantitative recovery and others can be eliminated. Under ideal circumstances the recovery rates of analyte and standard are identical, i.e. the same amount of both compounds is lost.

An internal standard should have properties as similar as possible to those of the analyte. This requirement can cover many properties, depending on the analytical problem: solubility, extraction behavior, unwanted (irreversible) adsorption, volatility, detector properties (such as absorption wavelength and extinction coefficient) and more. In addition it should be eluted preferably in the middle of the chromatogram or in proximity to the peaks of interest; very short or long retention times are unwanted.

The figure presents some excellent examples from the literature. Pronethalol is suitable as an internal standard for the analysis of propranolol, a beta blocker; 3,4-dihydroxybenzylamine is suitable for the catecholamines dopamine (which has only one additional CH_2 group) and epinephrine; vitamin D2 is suitable for the determination of vitamin D3 in cod-liver oil, a method which needs intricate sample preparation. Vitamin D2 has only one additional double bond and is not found in cod-liver oil. A perfect example is also shown in the section on the two-method verification of an analytical result ($\rightarrow$ 3.9).

pronethalol   for   propranolol

3,4-dihydroxy-   for   dopamine   and   epinephrine
benzylamine

vitamin D_2 for   vitamin D_3

3.5 A Linearity Test

In many but not all cases, a calibration curve is linear ($\rightarrow$ 2.52). A simple test for linearity is the investigation of the response factor, i.e. the ratio of signal to concentration as a function of the concentration. We use again the data of 2.52:

x concentration	y_1 signal non-linear case	y_1/x response non-linear case	y_2 signal linear case	y_2/x response linear case
1	1.2	1.20	1.2	1.20
2	2.1	1.05	1.9	0.95
3	3.0	1.00	3.0	1.00
4	3.9	0.98	4.1	1.03
5	4.8	0.96	4.8	0.96
6	5.9	0.98	6.1	1.02
7	7.0	1.00	7.0	1.00
8	8.1	1.01	7.9	0.99
9	9.2	1.02	9.2	1.02
10	10.3	1.03	9.7	0.97

Shown as graphs, it becomes clear that the first relationship follows a trend. Starting at 1.2, the response factor drops to a minimum in the middle of the interval before increasing steadily. Contrasting to this behavior, the response factor oscillates in the linear case and no trend can be observed. This oscillation is a proof of linearity although it can be of a more random type than in the example discussed here. With fewer data points it may be difficult to decide between linearity and non-linearity.

Note that the first data point 1/1.2 is dubious and seems to be out of the allowed range of the calibration function. It may be below the limit of quantitation.

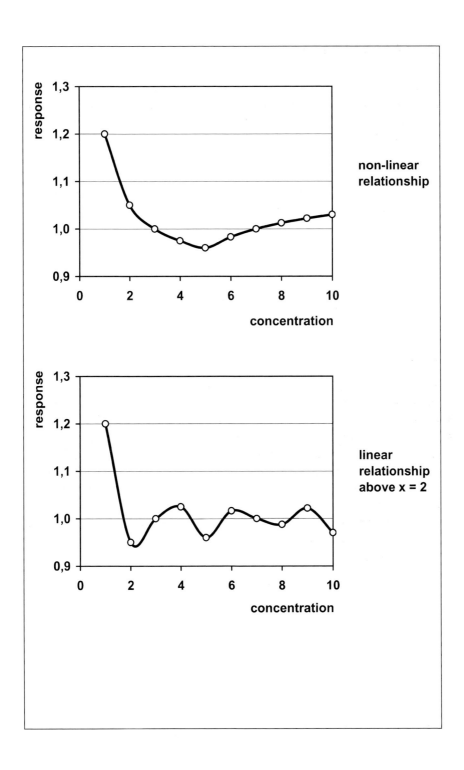

3.6 Rules for Accurate Quantitative Peak Size Determination

In quantitative chromatography several rules exist which may seem trivial and which quite often are not observed. Sometimes the analysts are satisfied with a method which separates the peaks of interest to a certain extent but not all of them to baseline. Then the accuracy of peak integration can be affected ($\rightarrow$ 2.43 to 2.48) but this will not be recognized by somebody who is mainly interested in precision ($\rightarrow$ 1.9).

Therefore the first rule requires baseline separation for all peaks which will be determined quantitatively. The necessary resolution increases with extreme peak size ratios: if resolution of, e.g., 1.5 is adequate for two peaks of equal size this value needs to be increased to, e.g., 3 for a pair with area ratio 100 : 1, depending on the extent of tailing.

Tailing reduces resolution and has a detrimental effect on the integration of small peaks behind large ones ($\rightarrow$ 2.45 to 2.47). Therefore the second rule demands the suppression of tailing by all possible means: minimum extra-column volumes in the instrument, excellent column packing quality, unhindered mass transfer between mobile and stationary phase (although it is not possible to give general recommendations), suitable sample solvent ($\rightarrow$ 2.9), and no column overload as a result of the injection of too much sample.

For small peaks the determination of peak height can be more accurate than that of peak area; at low signal-to-noise ratios it is also more precise ($\rightarrow$ 2.41). For peak height measurements it is, however, essential that constant retention factors are guaranteed ($\rightarrow$ 2.37). If it is not possible to reach the necessary resolution the small peak must be eluted before the large one and be quantitated by height ($\rightarrow$ 2.46). Only with high peak symmetry is the order of elution of no importance.

Accurate Quantitative Analysis

1. Baseline resolution

2. Avoid tailing

3. Height determination for small peaks

4. Small peaks must be eluted in front of large ones if baseline resolution is not possible

3.7 High–Low Chromatography

The analysis of sample purity in the 99% range and higher is prone to many sources of error if it is performed with the 100% method and a single sample concentration ($\rightarrow$ 2.51). It is better to perform this type of analysis with two different dilutions – at high and low concentration. The first determination is for the quantitation of the trace components whereas the second enables analysis of the main compound to be performed.

The two concentrations are chosen in such a way that the peaks of interest give peak heights of ca. 0.5 absorbance units. Therefore the dilution corresponds to the mass ratio of the main and trace peaks: if the impurities are in the 1% range one of the injection solutions must be diluted 100-fold compared to the other. To exclude any solvent effects ($\rightarrow$ 2.8 to 2.10) it is best to dissolve all samples in the mobile phase. The subsequent chromatography is then performed under completely identical conditions (including the sample volume) and consecutively; to avoid carry-over it is better to inject the low concentration first (otherwise the injector should be rinsed between the two analyses). The content of impurities is then obtained directly from the integration data corrected by the dilution factor.

With this method the only source of error is the quality of the dilution step. All other errors are compensated.

Analysis of a single concentration by changing the detector range is less favorable because the detector linearity needs to be investigated and because the signal-to-noise ratio is worse at lower ranges.

Chromatographic conditions:

Sample:	peptide
Column:	25 cm length
Stationary phase:	Zorbax 150 A C8, 5 µm (reversed phase C_8)
Mobile phase:	0.1 M sodium phosphate pH 2.6/acetonitrile 71 : 29
Detector:	UV 214 nm

Reference: E. L. Inman and H. J. Tenbarge: J. Chromatogr. Sci. 26 (1988) 89

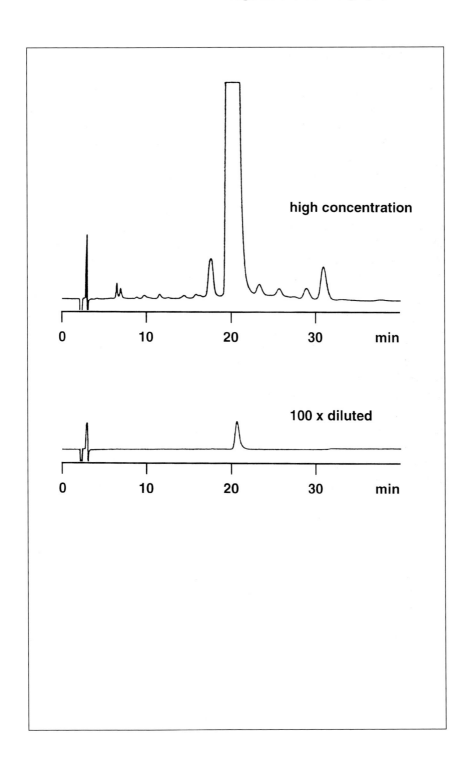

3.8 Control Charts

Control charts are a simple and effective tool for the documentation of analytical data, column test results, or any other determinations over a long period of time. The data are noted consecutively on a horizontal time axis; this can be done manually or by a computer. By this system it is possible to recognize trends or outliers at a glance, and perhaps to assign them to distinct events: Are the best results always obtained when a new bottle of solvent had been opened? Are poor results produced on Mondays? Upper and lower limits can be drawn which lead to the rejection of samples or to certain actions or preventive means. The limits can be based on statistics or on empirical experience. It is recommended to use limits of $\pm$ 3 standard deviations. With only random errors 99.7% of analytical results will be within $\pm$ 3 s ($\rightarrow$ 1.10). Data with a higher deviation are most probably subject to systematic error.

The example shown documents a procedure for polymer analysis in standard immunoglobulin preparations. The test was performed twice a day. The standard has an unwanted polymer content of ca. 0.8%. Outliers such as obtained on June 22 led to re-examination of the analytical method.

Chromatographic conditions:

Sample:	immunoglobulin preparation from blood, 200 µg protein
Column:	7.5 mm × 60 cm
Stationary phase:	TSK G3000SW, 10 µm (silica with hydrophilic phase for size exclusion chromatography)
Mobile phase:	phosphate buffer pH 7.0, 0.5 mL min^{-1}
Detector:	UV 280 nm
Test:	content of immunoglobulin polymers, calculated as peak area percent
	P = polymers, D = dimers, M = monomers

Reference: R. Brügger, Central Laboratory, Blood Transfusion Service, Swiss Red Cross, Bern, Switzerland

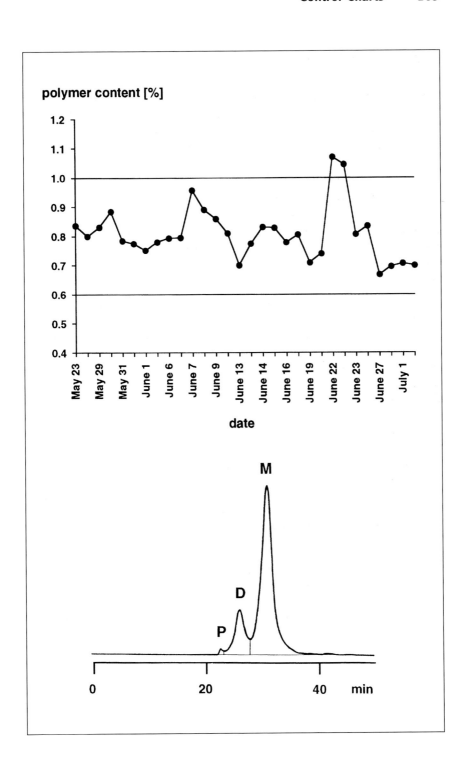

3.9 Verification of the Analytical Result by Use of a Second Method

The best technique for the verification of an analysis is to investigate the sample by a second method. It is a matter of course that both procedures are calibrated correctly by use of totally independent multi-point calibration curves. The two methods should be as different as possible: ion chromatography and atom spectrometry, liquid and gas chromatography, exclusion and reversed-phase chromatography. If both methods yield identical results this is almost proof of accuracy.

For the analysis of toxic amines in fried meat it was not possible to verify the results other than by using two different reversed-phase HPLC methods. In the case of deviations a traceable analytical error could always be found retrospectively, which means, on the other hand, and with high probability, that matching results are accurate. Without a second method, however, it would be impossible to recognize the erroneous analyses.

Chromatographic conditions:

Sample:	standard mixture of carcinogenic amines which are found in meat fried at too high a temperature, with internal standard
Column:	4.0 mm × 25 cm
Stationary phase:	LiChrospher 60 RP-Select B, 5 μm (reversed phase C_8)
Method 1:	A = 0.01 M triethylamine + phosphoric acid pH 3.3, B = acetonitrile. 5% B at 0–4 min, linear gradient to 25% B at 4–24 min
Method 2:	A = 0.05 M acetate buffer pH 4.7/methanol 9 : 1, B = acetonitrile/methanol 9 : 1. 3.6% B at 0–4 min, linear gradient to 27% B at 4–30 min
Detector:	UV 264 nm; for peak 6 UV 320 nm

Reference: P. Rhyn, O. Zoller and B. Zimmerli, Swiss Federal Office of Public Health, Section of Food Chemistry, Bern, Switzerland

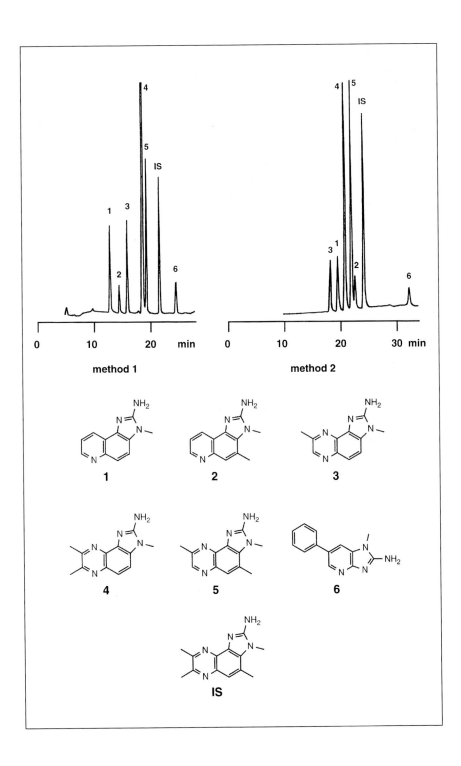

3.10 Description of Ruggedness

When passing on a method ($\rightarrow$ 3.11) it is especially important to draw the attention to the parameters with poor ruggedness ($\rightarrow$ 1.13). This enables special regard to be paid to these points when the method is repeated. If the procedure has been written carefully one can assume that other parameters are not critical (if a full investigation was not possible it should be stated that the ruggedness of the one or other of the parameters is not known).

An example from a description of a method for the quantitative analysis of gentamicin in milk shows clearly how critical ruggedness should be presented (faithful quotation):

"Interferences: Of the common veterinary drugs (other aminogly-cosides, penicillins, tetracyclines, macrolides, sulfa drugs, chloramphenicol, and novobiocin), neomycin is the only compound that will be detected if present in milk. Partial peak overlap is observed."

"The methanol content of the mobile phase is critical; a deviation of 1–2 % causes 5–10 minute shifts in the retention time."

"The necessary amount of methanol can vary from column to column, even when columns from the same manufacturer are used. The retention time decreases as the column ages. The user will need to decrease the amount of methanol periodically, however not daily."

Chromatographic conditions:

Sample:	extract of milk from a cow which was treated with gentamicin (gentamicin has four components)
Column:	4.6 mm $\times$ 15 cm
Stationary phase:	Spherisorb ODS2, 5 µm (reversed phase C_{18})
Mobile phase:	5.6 mM sodium sulfate with 11 mM sodium pentane-sulfonate and 0.1 % acetic acid/methanol 82 : 18, 1.5 mL min^{-1}
Temperature:	32.5 °C
Detector:	post-column derivatization with ortho-phthaldialdehyde at 33 °C, then fluorescence 340/430 nm

Reference: P. J. Kijak, J. Jackson and B. Shaikh: J. Chromatogr. B 691 (1997) 377

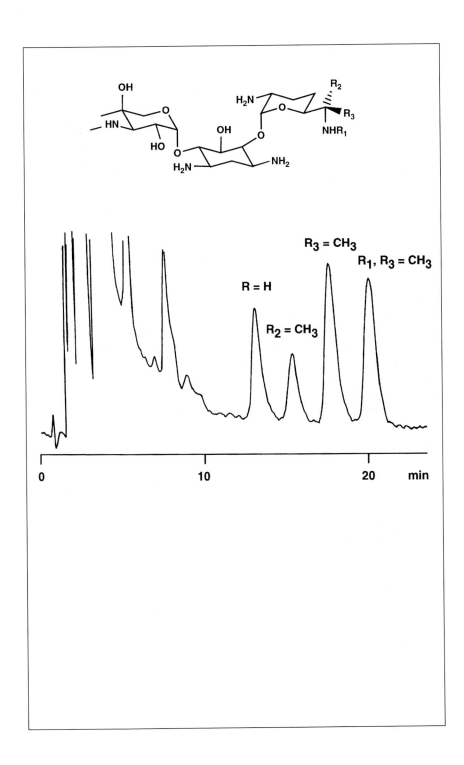

3.11 Rules for Passing On an HPLC Method

It is necessary to note in detail very many procedures and data if one wants to use a method in another laboratory. This applies to internal methods within a company as well as to scientific publications.

A problem on its own is always the sampling procedure. It is also possible that sample storage and preparation will influence the analytical result. These need detailed descriptions, including the preparation of the injection solution ($\rightarrow$ 2.7 to 2.9, 2.12 to 2.14) as the last step.

The preparation of the mobile phase must be described in detail: solvent mixing ($\rightarrow$ 2.1), quality and amount of reagents ($\rightarrow$ 2.4, 2.5), pH ($\rightarrow$ 2.2, 2.3). The temperature of the column should also be stated ($\rightarrow$ 2.29) (otherwise add the remark 'separation at ambient temperature, the influence of temperature was not investigated').

Extra-column volumes ($\rightarrow$ 2.20) and, for gradient separations, the dwell volume ($\rightarrow$ 2.21) are part of a method description.

Column dimensions and the precise name of the stationary phase ($\rightarrow$ 2.23, 2.24), in extreme cases even the batch ($\rightarrow$ 2.25), need to be given as a matter of course as well as the volume flow rate ($\rightarrow$ 2.31, 2.38) and, if applicable, the gradient run time ($\rightarrow$ 2.32).

If a UV or fluorescence detector is used the wavelength must be known ($\rightarrow$ 2.33, 2.34) and detector calibration ($\rightarrow$ 3.3) at regular time intervals is a prerequisite.

The integrator parameters can influence data acquisition. With data systems which allow re-integration of the chromatogram this is less important.

Unfortunately untoward surprises are almost to be expected if a method comes from academia because university institutes rarely install a formal quality assurance system ($\rightarrow$ 3.18).

Specifications which need to be known for successful method transfer

Sampling, sample storage, sample preparation

Sample solvent

Mobile phase preparation, reagent quality

Temperature

Extra-column volumes

Dwell volume of gradient separations

Column dimensions

Volume flow rate of the mobile phase

Detailed identity of the stationary phase

Detector data

Integration parameters

3.12 Quality Assurance in the Laboratory

For every analytical activity in the laboratory it is worthwhile working in accordance with a quality assurance system such as ISO 9001 or EN ISO/IEC 17025 ($\rightarrow$ 3.18). At first glance such a system might seem to be a list of irksome instructions; it is, in fact, a security net under the tightrope of accuracy and precision which prevents headlong falls.

Possible requirements for a quality assurance system for HPLC are listed on the opposite page. Depending on the tasks of the laboratory it might be longer. Most actions are trivial but the personnel needs to understand their necessity and to follow them meticulously.

Inadequate ($\rightarrow$ 2.4, 2.5, 2.12) or ill-defined quality of reagents or solvents gives rise to troubles and unsatisfactory precision. Because the decomposition or contamination of a chemical can begin after the first opening of a bottle it is advisable to note both the date of opening and the date after which the chemical should no longer be used on the label. It is well possible that a reagent can still be used for a less demanding application after this time limit.

Columns need to be tested before their first use and later again at regular intervals ($\rightarrow$ 3.1). It is necessary to define binding criteria such as theoretical plate number, asymmetry, and pressure which lead to the replacement of a column; the old one is then sent back for re-filling and not kept in a locker. The test chromatograms or the computer raw data are stored in such a way that they can be easily consulted. It is best to keep a file which contains all the relevant data for each column.

As for columns, it is necessary to test the instruments at specified intervals ($\rightarrow$ 3.2). Each apparatus should have its log-book where the test results and the necessary services are noted. A log-book is kept in close proximity to its instrument.

All kinds of data can be presented clearly in the form of control charts ($\rightarrow$ 3.8).

An important part of quality control, though not discussed here, is obviously the faithful documentation of the analytical data.

Quality Assurance in the Laboratory

Use chemicals of well-defined quality

Note the opening date of reagent bottles. Define the maximum time-span of use

Perform column tests

Repeat column tests at regular intervals

Define column specifications

Keep log-books for the columns

Perform apparatus tests

Repeat apparatus tests at regular intervals

Define instrument specifications

Keep log-books for the instruments

Use control charts

3.13 Standard Operating Procedures

A standard operating procedure or SOP consists of detailed and written instructions for performing an activity in the laboratory. This may be a procedure for weighing or procedures for more complex tasks such as a demanding sample preparation or the determination of a calibration curve. Apparatus tests ($\rightarrow$ 3.2) are also described in the form of SOPs. If an SOP is given, it should be possible to perform a method in another laboratory by competent yet not specially trained personnel ($\rightarrow$ 3.11) because the SOP should include all necessary steps of the work in appropriate detail. The SOP can either refer to a certain procedure, e.g. 'Ion Chromatographic Determination of Lead in Vegetables' or, in a more general manner, to all possible samples which could be of interest for lead analyses.

To set up a collection of SOPs in a laboratory it is best they be written by the persons who perform the activity, i.e. not by their superiors. The latter, however, need to control the completeness of the description; no working step must be omitted because it seems too obvious. The SOPs are then collected in files which are at hand in the laboratory; they must not disappear into an office. It is a matter of course that SOPs are dated and that they can be revised. The most recently updated and numbered version is in use whereas all older versions are kept in an archive.

SOPs can include specifications of the test or analysis which need to be fulfilled if a sample, a calibration curve, or an instrument is to be regarded as satisfactory. (This is not true, of course, for pure descriptions of a task such as how to write the laboratory notebook.) Following SOPs is mandatory! If not, they are not worth the effort expended writing them.

Standard Operating Procedures

Each activity in the laboratory is described by an SOP, especially apparatus tests and methods of analysis.

SOPs are detailed descriptions of a particular activity which enable it to be performed accurately by a competent person without special training.

An SOP is written by the person who actually performs the given activity.

The collection of SOPs is kept in the laboratory.

SOPs may include specifications of the test or analysis which need to be fulfilled.

Following SOPs is mandatory.

3.14 Method Validation

Method validation is the process which proves that an analytical procedure is suited for the intended purpose. Only with validated methods is there a guarantee that the laboratory produces accurate and precise analytical results. Lucky hits and trimmed data are not reconcilable with validation.

A prerequisite for validation is instruments which have passed the apparatus test ($\rightarrow$ 3.2). All software which is used needs to be designed such that results cannot be altered and no files can disappear. The laboratory staff works according to the rules of quality assurance ($\rightarrow$ 3.12). Only Standard Operating Procedures (SOPs) are used as methods of analysis ($\rightarrow$ 3.13).

Validation is the knowledge and documentation of the selectivity (perhaps specifity), range, linearity ($\rightarrow$ 2.52, 3.5), quantitation limit, detection limit, precision, accuracy ($\rightarrow$ 1.9), and ruggedness ($\rightarrow$ 1.13) of a certain procedure. The corresponding definitions can be found on the opposite page.

Depending on the analytical problem it is perhaps not necessary to determine all these parameters (although this is the rule). If, e.g., the main product which constitutes 10 to 90% of a formulation is to be analyzed, one is not interested in the quantitation limit.

Control charts ($\rightarrow$ 3.8) can be extremely helpful during routine use of a method. If the chart shows that the analytical results are free from outliers for a long period of time it might be possible that multi-point calibration is not necessary because linearity is not a problem. Under such conditions the laboratory can switch to one-point calibration but the personnel will check the results by the continuous use of the control chart.

Validated Procedures

Validation means knowledge and documentation of

Selectivity: The compound of interest can also be identified and quantified in the presence of other (similar) compounds.

Specifity: Only the compound of interest is determined whereas other components or features of the sample do not influence the result.

Range: Range of concentration or mass for which the procedure can be used with appropriate precision and accuracy.

Linearity: Straight calibration curve, i.e. the data points are directly proportional to concentration or mass.

Quantitation limit: Lowest concentration or mass which can be quantitated precisely and accurately (often 10x noise level).

Detection limit: Lowest concentration or mass which can be detected qualitatively (often 3x noise level).

Precision: Standard deviation. Differentiate between repeatability (same column, same instrument, same laboratory...) and reproducibility (not same laboratory...).

Accuracy: Difference between found and true value.

Ruggedness: Insensitivity to small fluctuations of the parameters of the procedure.

3.15 Some Elements of Validation

Some elements of validation which are discussed on the previous page ($\rightarrow$ 3.14) can be drawn into the presentation of a calibration curve ($\rightarrow$ 1.14) for the purpose of visualization.

Limits of quantitation and detection: Here the detection limit with approx. threefold noise level is drawn. A peak at the quantitation limit should have a height of approx. ten times the noise level ($\rightarrow$ 1.17).

"True value" (black point): The true value of an analysis, i.e. the true concentration of analyte, is unknown by definition. Besides the fact that it would be necessary to perform a large number of measurements in order to get a reliable mean the analyst can never be sure about the absence of unknown and uncontrollable systematic effects which could affect the result.

Trueness: The difference between the found and the true value. Since the true value is not known, the trueness is unknown as well.

Repeatability (grey points): In the situation shown here the repeatability is the standard deviation of several peak sizes (area or height) which were obtained by the multiple injection of a certain sample solution.

Linearity (white points): The distribution of the data found with the calibration runs shows that it is justified to postulate a linear relationship between the concentration of the analyte and the peak size. The points are dispersed in a random manner above or below the regression line ($\rightarrow$ 3.5).

Specificity: The detail from the chromatogram shows that the analyte peak seems to be free from interfering peaks. During the validation it is necessary to investigate by diode array detection, mass spectroscopy and/or another analytical method ($\rightarrow$ 3.9) if a coelution can be excluded with high probability.

Reference: J. Ermer, Aventis Pharma, Frankfurt/Main, Germany

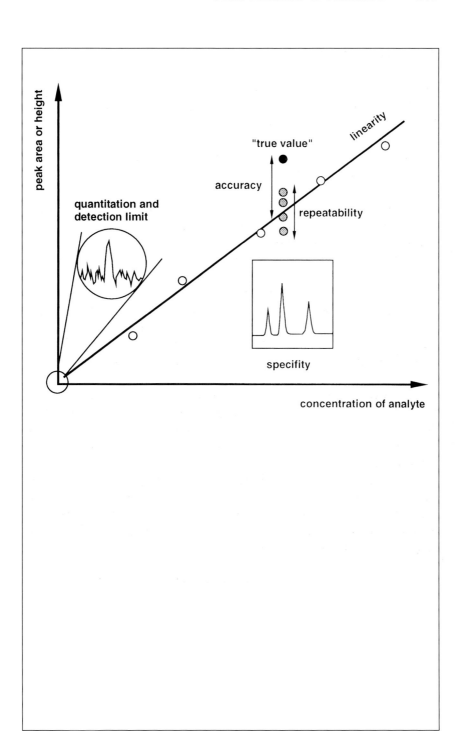

3.16 A Validation Example

This Section presents a practical example of method validation. The goal is the quantitative determination of 4-amino-3-hydroxybutyric acid (γ-amino-β-hydroxybutyric acid, GABOB) in pharmaceutical tablets.

Chromatographic conditions: Column: 4 mm × 25 cm; stationary phase: Purospher C_{18}, 5 µm; mobile phase: water with 10 mM sodium heptanesulphonate, pH 2.4, 1 mL min^{-1} temperature: 40°C; detection: UV 210 nm.

Specificity: The formation of degradation products was induced by the 18-h treatment of a tablet at 80°C. A placebo tablet with 5-ethyl-5-phenylbarbituric acid, pyridoxine, phenytoin, microcrystalline cellulose, and magnesium stearate was also analyzed.

Stability of solutions: Solutions with 0.5 mg mL^{-1} GABOB (plain and spiked to a placebo) were stored for 6 days in a refrigerator.

Robustness: The influence of mobile phase pH was checked at pH 2.0, 2.4, and 3.0, and the counter-ion concentration was changed from 9 mM to 10 mM and 11 mM.

Linearity: The usual analyte concentration in the sample solution is 0.5 mg mL^{-1}, therefore the linearity was investigated in the range from 0.4 mg mL^{-1} to 0.6 mg mL^{-1} with 5 points (3 replicate injections each).

Accuracy: Known amounts of analyte were spiked to a placebo sample, giving concentrations of 0.4, 0.5 and 0.6 mg mL^{-1} (3 replicate injections each).

Precision, repeatability: The assay was identical with the one for accuracy. In addition, 3 standard solutions of 0.4, 0.5 and 0.6 mg mL^{-1} were prepared for calibration.

Precision, intermediate precision: The assay was identical with the one for accuracy but all measurements were also performed on a HPLC instrument of another manufacturer. All other parameters, including the column, were identical.

Range: As mentioned above.

Reference: M. Candela, A. Ruiz, F.J. Feo: J. Chromatogr. A 890 (2000) 273
See also: United States Pharmacopoeia (USP) 28 (2005), Section 1225,
p. 2748, Validation of Compendial Methods

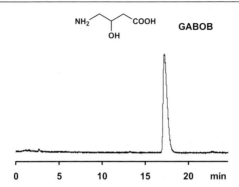

GABOB

Specificity: The analyte peak was still well resolved and its purity (by the diode array detector peak purity function) was > 99.0 %. The placebo tablet did not show any peak at the retention time of GABOB.

Stability of solutions: No decomposition was observed, and peak areas were unchanged.

Robustness: The pH has a strong influence on retention factors (extremes of $k = 8.3$ to $k = 12.8$) and peak shape whereas the counter-ion concentration has not.

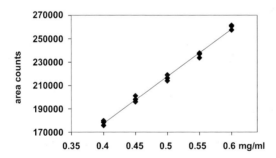

Linearity: Calibration function: $y = 402430\ x + 16491$. The mean relative standard deviation of the response factors was 1.7 %, and the correlation coefficient was 0.997.

Accuracy: The recoveries found varied between 99.30 % and 100.82 % (mean of the 9 values = 100.1 % ± 0.4 %).

Precision, repeatability: Relative standard deviation of the normalized experimentally determind concentrations = 0.68 %.

Precision, intermdediate precision: Instrument A gave a relative standard deviation of the normalized experimentally determined concentrations of 0.68 % and a mean of 1.0008. Instrument B gave 1.23 % and 0.9874, respectively.

Range: The assay is suitable in the concentration range 0.4–0.6 mg mL^{-1} for tablets with 100 mg GABOB.

3.17 Measurement Uncertainty

The document "Quantifying Uncertainty in Analytical Measurement" gives a useful (and in some cases mandatory) guideline on how to determine the measurement uncertainty of a quantitative chemical analysis. The schematic procedure is presented in the graph.

Specify measurand: The standard operating procedure ($\rightarrow$ 3.13) must be understood. (If not, write it in a clearer and more comprehensible style or educate the personnel.) The complete equation of the measurand must be known. In the case of a one-point calibration HPLC analysis without sample preparation it is of the type:

$$c = \frac{A_s}{A_r} \cdot \frac{m_r \cdot P_r}{V_r} \cdot \frac{V_s}{m_s}$$

with c: concentration of the analyte in the sample (measurand); A: peak area; m: mass; P: purity; V: volume of measuring flask; s refers to the sample and r to the reference.

Identify the uncertainty sources: An Ishikawa diagram can be very helpful ($\rightarrow$ 1.18). It is strictly based on the equation of the measurand.

Quantify the uncertainty sources: Look for experimental data (such as repeatabilities, validation data) and/or published data (such as purities, technical data of the glassware and pipettes). Transfer the data into standard uncertainties.

Calculate the combined standard uncertainty u_c ($\rightarrow$ 1.11): For the equation given above it is calculated as follows (as relative value):

$$\frac{u_c(c)}{c} = \sqrt{\begin{array}{l} \left(\frac{u(A_s)}{A_s}\right)^2 + \left(\frac{u(A_r)}{A_r}\right)^2 + \left(\frac{u(m_r)}{m_r}\right)^2 + \left(\frac{u(P_r)}{P_r}\right)^2 \\ + \left(\frac{u(V_r)}{V_r}\right)^2 + \left(\frac{u(V_s)}{V_s}\right)^2 + \left(\frac{u(m_s)}{m_s}\right)^2 \end{array}}$$

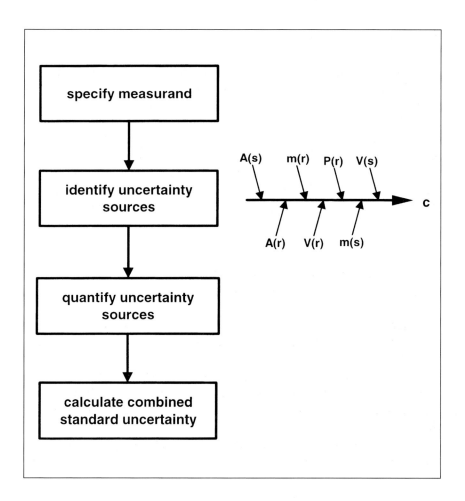

It is then necessary to critically review the calculated value. It may be indicated to re-evaluate some uncertainty sources, especially the large ones or the ones with poor reliability. In order to decrease the combined measurement uncertainty it is essential to improve the sources of large uncertainty (e.g. a standard of poor purity).

Reference: S. L. R. Ellison, M. Rösslein and A. Williams: Eurachem/Citac Guide "Quantifying Uncertainty in Analytical Measurement", 2nd edition (2000). Free download: http://www.measurementuncertainty.org/mu/guide.

3.18 Formal Quality Assurance Systems

Validated methods are the prerequisite for the establishment of a formal quality assurance system. Several such systems are in use:

GLP (Good Laboratory Practice) is the oldest. It was developed in the USA as an answer to cases of scientific fraud in the drug industry. GLP includes an immense amount of paper work and guarantees that all necessary experiments were carefully performed; there is, however, no evidence of competence. GLP is essential for research laboratories in the pharmaceutical industry but it cannot be recommended for other enterprises.

EN ISO/IEC 17025 (European Norm) is the European system of accreditation for competence. It can be recommended to all laboratories whose services have external effect. Whoever gives a commission to an accredited laboratory can trust that the analyses are performed in a competent manner according to the definition on the opposite page. It is not necessary to repeat the analysis in one's own laboratory, and thus EN ISO/IEC 17025 helps to avoid multiple investigations. The system allows much more freedom than is often assumed: competence also includes the ability to reveal which analyses or tests are unnecessary.

ISO 9001 (International Standards Organization) represents a formal quality management system which has been developed mainly for the consumer goods industry (e.g., for the automobile industry). It defines mandatory procedures but gives no proof of competence. Today it is almost essential for a company to be certified according to ISO 9001; if its laboratory has an EN ISO/IEC 17025 accreditation with greater demands it will fit smoothly into the ISO 9001 system.

A laboratory or company which wants to obtain accreditation or certification has to undergo a thorough examination by external experts, which later will be repeated regularly. The effort is rewarded by the certainty that the analytical results are precise and accurate.

Formal Quality Assurance Systems

GLP (Good Laboratory Practice):
for the pharmaceutical industry,
prevention of fraud

EN ISO/IEC 17025 (European Norm):
for testing facilities,
proof of competence

> Qualilfied personnel perform well-defined tests in
> accordance with validated methods and with
> tested measurement equipment. All Prodecures
> are clearly regulated and adequate documenta-
> tion is guaranteed.

"Methods" are the Standard Operating Procedures
"Measurement equipment" is the instruments for analysis
"Tests" are the analyses

ISO 9001 (International Standards Organization):
for the consumer goods industry,
formal quality management system,
process-oriented

Index